D1207119

ENCYCLOPEDIA OF TIMBER FRAMING AND CARPENTRY

The Modern Carpenter Joiner and Cabinet-Maker

ENCYCLOPEDIA OF TIMBER FRAMING AND CARPENTRY

G. Lister Sutcliffe, Editor
Associate of the Royal Institute of British Architects, member of the Sanitary Institute, editor and joint-author of *Modern House Construction*, author of *Concrete: Its Nature and Uses*

Roy Underhill, Consultant
Television host of "The Woodwright's Shop", author of *The Woodwright's Shop*, *The Woodwright's Companion*, and *The Woodwright's Workbook*, and Master Housewright at Colonial Williamsburg

A Publication of
THE NATIONAL HISTORICAL SOCIETY

The Modern Carpenter Joiner and Cabinet-Maker presented up-to-date techniques and tools for its time. However, much has changed since 1902. Not all materials and methods described in these pages are suitable for the construction materials and tools of today. Before undertaking any of the building, remodeling or other practices described in these pages, the reader should consult with a reputable professional contractor or builder, especially in cases where structural materials may come under stress and where structural failure could result in personal injury or property damage. The National Historical Society, Cowles Magazines, Inc., and Cowles Media Company accept no liability or responsibility for any injury or loss that might result from the use of methods or materials as described herein, or from the reader's failure to obtain expert professional advice.

Library of Congress Cataloging-in-Publication Data
Encyclopedia of timber framing & carpentry / G. Lister Sutcliffe, editor; Roy Underhill, consultant.
p. cm. — (The Modern carpenter joiner and cabinet-maker)
Reprint. Originally published: England, 1902.
ISBN 0-918678-58-7
1. Wooden-frame houses. 2. Carpentry. I. Sutcliffe, G. Lister. II. Underhill, Roy. III. Title: Encyclopedia of timber framing and carpentry. IV. Series.
TH4818.W6E53 1990
694—dc20 90-6345
CIP

CONTENTS

DIVISIONAL-VOL. IV

SECTION VII.—CARPENTRY

BY THE EDITOR

ILLUSTRATIONS

DIVISIONAL-VOL. IV

ILLUSTRATIONS IN TEXT

SECTION VI

THE STRENGTH OF TIMBER AND TIMBER FRAMING

SECTION VII.—CARPENTRY

PREFACE

"Science is a first-rate piece of furniture for a man's upper chamber,
if he has common sense on the ground floor."
Oliver Wendell Holmes, 1858

As a physician, Oliver Wendell Holmes well understood the difficulty of imposing mathematical predictability upon the messy organic world. His most familiar bit of popular verse draws its humor from just the problem attacked in this volume, that of measuring and predicting the behavior of infinitely variable wood. The subject of his "Logical Story," "the Deacon's Masterpiece" was a carriage built with every part as strong as the others, so that it could never break down, but only wear out. The Deacon, of course, started with first-class timber.

"So the Deacon inquired of the village folk
Where he could find the strongest oak,
That couldn't be split nor bent nor broke,"

And by using only the best wood, including ash "from the straightest trees," his carriage ran for one hundred years to a day and then —

"it went to pieces all at once,
All at once, and nothing first,
Just as bubbles do when they burst."

This fond hope is based on a fundamental quest of the builder. Joseph Moxon stated both the objective and the obstacle of this pursuit in his 1678 *Mechanic Exercises.* In his lessons on making the basic mortice and tenon joint, he observed that "if one be weaker than the other, the weakest will give way to the strongest when an equal violence is offer'd to both. Therefore you may see a necessity of equallizing the strength of one to the other, as near as you can. But because no rule is extant to do it by, nor can (for many Considerations, I think,) be made, therefore this equallizing of strength must be referred to the Judgement of the Operator."

It's the old "a chain is only as strong as its weakest link" problem. But not even the variability of wood could damp the insufferable human drive to measure and quantify, although wise nineteenth-century men of science, such as the great Peter Nicholson, knew the limits of their scientific testing. "Yet it is impossible to account for knots, cross-grained wood, &c., such pieces being not so strong as those which have straight fibres; and if care be not taken in choosing timber for a building, so that the fibres be disposed in parallel straight lines, all rules which can be laid down will be useless."

One problem with strength tables is the changing nature of the material, as the old, slow-grown timber is replaced by faster grown, knottier, and coarser stuff. In 1786, the Carpenter's Company of Philadelphia tried to address this problem in the only way they knew, by raising prices. Their Rule Book (a secret price-fixing manual) complained that "The stuff also used at this time is certainly from one sixth to one eighth more labour than that used some years ago, it being in general so much worse — and to expect work now, ... for the same price by the square that the workmen had then, can hardly be deemed equitable or just."

It's still the same old story. Aside from raising prices, builders have constantly had to adapt their methods to meet the available labor, skills, and materials. Although written almost four centuries ago in 1603, the following passage could come from the operator of any computer-assisted truss design and manufacturing plant today. "The careless waste... of our wonted plenty of timber, and other building stuffe, hath enforced the witt of this latter age to devise a new kind of compacting, uniting coupling, framing, and building, with almost half the timber which was wont to be used, and far stronger."

Even the same piece of timber relentlessly changes in character from the moment when it is first cut down. To further complicate matters, the strength of a structure is determined by both the greenness of the wood, as well as the greenness of the carpenter. An off-square saw cut on the base of a column can reduce its strength by forty percent. One builder working on the College of William and Mary in 1704 was hauled into court because "partly by the Plank & timber being green and unseasoned & partly by employing a great number of unskilled workmen to complying his haste, [the building] was shamefully spoilt."

So here you have the tables to calculate the strength of timbers and joints. This information, coupled with a generous safety factor, should allow you to design structures to meet the challenge of the centuries. But no table or formula can be trusted without an equally trustworthy and experienced eye for the quality of wood and workmanship. Oliver Wendell Holmes put it best. "Knowledge and timber shouldn't be much used till they are seasoned."

ROY UNDERHILL
MASTER HOUSEWRIGHT
COLONIAL WILLIAMSBURG

SECTION VI

THE STRENGTH OF TIMBER AND TIMBER FRAMING

PART I.—THEORETICAL PART II.—PRACTICAL

BY

THE EDITOR

SECTION VI

THE STRENGTH OF TIMBER AND TIMBER FRAMING

PART I.—THEORETICAL

CHAPTER I

STRESSES AND STRAINS

Definition.—The terms stress and strain are often confused.

It is best to consider the term "stress" to mean the force applied to a body, and "strain" to be the internal disturbance in that body which the force makes or tends to make. Thus, in the case of wood, a tensile stress applied in the direction of the grain causes a strain, which, if severe enough, will result in the elongation and rupture of the fibres, as a piece of elastic is broken if pulled beyond its limit of elasticity. But this strain, although caused by a purely tensile *stress*, is not wholly a tensile strain; it is in part a compressive strain at right angles thereto. If a piece of elastic is pulled, its length increases but its diameter gradually decreases; in other words, while its fibres are elongated in the direction of the *stress* they are compressed in a direction at right angles thereto. Similarly, a piece of wood subjected to a compressive stress develops a tensile strain at right angles to the stress as well as a crushing strain in the direction of the stress.

The principal stresses to which timber is subjected in carpentry are: (1) Tension, (2) Shearing, (3) Compression, and (4) Transverse Stress or Bending. Torsion or twisting need not be considered.

1. *Tension.*—A bar is said to be subjected to tension, or to a tensile stress, when two opposite forces are applied to it in such a manner as to tend to elongate or rupture it by direct pulling. In carpentry the most common example is the tie-beam of a roof, which is subjected to a tensile stress by the outward thrust of the principal rafters, one at each end of the beam.

2. *Shearing.*—When a force is applied to a bar in such a manner as to tend to cut the fibres across or to cause them to slide upon each other, it is called a shearing force or stress. Thus, in fig. 514 the foot of the principal rafter tends to cause the fibres of the tie-beam to slide upon each other along the plane *a b c*. This is known as "shearing with the grain", or "detrusion"; shearing *across* the grain may occur in trenails and tenons.

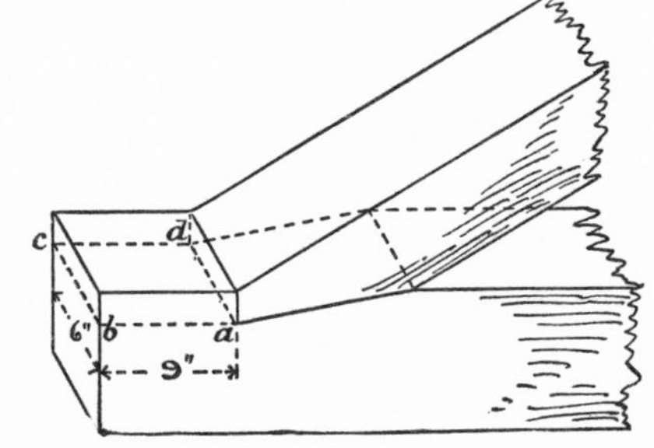

Fig. 514.—Shearing Stress in Tie-beam

3. *Compression.*—This is a squeezing or crushing stress, and tends to injure the bar by crushing the fibres. The struts of roof-trusses are examples of timbers subjected to this stress; so also are posts or columns supporting loads.

4. *Transverse Stress.*—This is a stress applied transversely to a bar, and tending to bend it and to break it across. Floor beams and joists are simple examples. It gives

rise to tensile and compressive strains, with shearing in addition. Strictly speaking, therefore, there are only three kinds of strains, namely, tension, compression, and shearing. In carpentry, timbers are frequently subjected to compound stresses. If the tie-beam of a roof supports a ceiling or floor it is subjected to both tensile and transverse stresses. A post very long in proportion to its diameter will fail by bending and transverse rupture, and not by simple crushing.

Stresses are usually expressed in lbs. per square inch. Thus, if a wrought-iron rod 1 inch square is pulled with a force of one ton (2240 lbs.), it is said to be subjected to a tensile stress of 2240 lbs. per square inch.

Strains, however, are measured by the ratio of distortion of the body in which they are set up. Thus, if the rod just mentioned is 12 inches long, and stretches $\frac{1}{1000}$ inch under the tensile stress of 2240 lbs., the tensile strain will be $\frac{1}{1000} \div 12 = \frac{1}{12000}$ per unit of length. That is to say, a rod of this material will be elongated $\frac{1}{12000}$ part of its length under a tensile stress of 2240 lbs.

Up to a certain point stress and strain are proportional. The same rod of iron under a stress of 2 tons would be lengthened $\frac{2}{1000}$ inch, under one of 3 tons $\frac{3}{1000}$ inch, and so on up to (say) 12 tons. Beyond this point, which of course varies for different materials and for different qualities of the same material, the strain increases more rapidly than the stress. The highest stress at which the normal proportion between stress and strain holds good is known as the *limit of elasticity* of the material. Up to this point the material is said to be perfectly elastic. When the stress is removed, the rod will spring back to its original length. Stresses greater than the limit of elasticity result in a permanent elongation of the material; when the stress is removed, the rod does not quite return to its original length. The difference between this original length and the length to which the rod returns when the stress is removed, is known as the *permanent set.*

The *Modulus of Elasticity* of any material is calculated from the stress and strain, provided that the limit of elasticity is not exceeded. It is expressed in lbs. per square inch, thus—

$$\text{Modulus of Elasticity (in lbs. per square inch)} = \frac{\text{Stress (in lbs. per square inch)}}{\text{Strain (per unit of length)}}.$$

In the case of the rod of wrought iron we have—

$$\text{Modulus of Tensile Elasticity} = 2240 \div \tfrac{1}{12000} = 26{,}880{,}000 \text{ lbs. per square inch.}$$

The same result is obtained if the calculation is made from any other stress and strain within the limit of elasticity; thus, if we take a stress of 3 tons and a strain of $\frac{3}{12000}$, we get—

$$\text{Modulus of Tensile Elasticity} = 6720 \div \tfrac{3}{12000} = 26{,}880{,}000 \text{ lbs. per square inch.}$$

In a similar manner the *Modulus of Compressive Elasticity* can be calculated from the shortening of a bar under a compressive stress, provided the limit of compressive elasticity is not exceeded.

The Modulus of Transverse Elasticity is calculated from the deflection of loaded beams, and will be considered, together with the *Modulus of Rupture*, in the chapter on beams.

The *Resistance*, or *Ultimate Strength*, or *Breaking Strength* of a material is the stress under which fracture occurs, and is usually expressed in lbs. or tons per square inch. Thus, the tensile resistance of wrought iron is described as being from 20 to 24 tons per square inch, and the compressive resistance of pitch pine about 4000 or 5000 lbs. per square inch.

The *Working Stress* of any material is the greatest stress which can safely be applied to it. It is sometimes known as the *Safe Load*, and is usually calculated by dividing the ultimate strength of the material by a divisor called the *Factor of Safety*. Thus, if the ultimate tensile strength of wrought iron is $22\frac{1}{2}$ tons per square inch, the working stress (with a factor of safety of 5) will be $4\frac{1}{2}$ tons per square inch.

Loads are of two kinds, "dead" and "live".

A dead load is, as the term implies, a still load, and is in many cases simply the weight of the structure itself. The dead load on a roof is the weight of the roof *plus* an allowance for snow; if a ceiling or floor is carried by the roof-timbers, the dead weight of these must be added to the weight of the roof and snow.

A live load is a moving load, and includes human beings, rolling stock and vehicles of all kinds, &c. Live loads are more severe in their effects than dead loads, and are often considered as producing a strain equal to a dead load of twice the amount. Thus, a man weighing $1\frac{1}{2}$ cwt. would be equivalent to 3 cwts. of dead load.

Repeated applications of a load reduce the strength of the structure to which the load is applied. Even a constant load continued for a long period has a similar effect. This weakening is known as the *fatigue* of the material. Allowance for it is usually made by adopting a sufficiently large factor of safety.

Impact is the term given to a load suddenly applied, such as a railway-train in rapid motion, or a body falling upon a floor or other structure. In warehouses, bridges, &c., allowance must be made for loads of this kind, as the momentary deflection caused by them is very great, and as they induce fatigue in the members of the structure.

CHAPTER II

FORCES AND MOMENTS

Force.—Numerous definitions of force have been proposed; the following is based on those of Professors Lanza and Rankine:—

Force has a tendency to change the relative rest or motion of the two bodies between which the tendency exists.

Forces, therefore, do not necessarily produce motion, as they may be resisted by equal and opposite forces. Thus, a load suspended on a cord is a downward force which is resisted by the equal and opposite upward force exerted by the cord.

A force has three characteristics—namely, *Point of Application, Direction, and Magnitude*—all of which can be represented by a single straight line, as in fig. 515, where a central load of 500 lbs. on the beam C D is represented by the straight line A B. The length of the line is proportional to the weight (according to the scale of lbs. adopted); the direction of the force is indicated by the arrow at B, and the point of application is at B. When these three characteristics are known, the force is known.

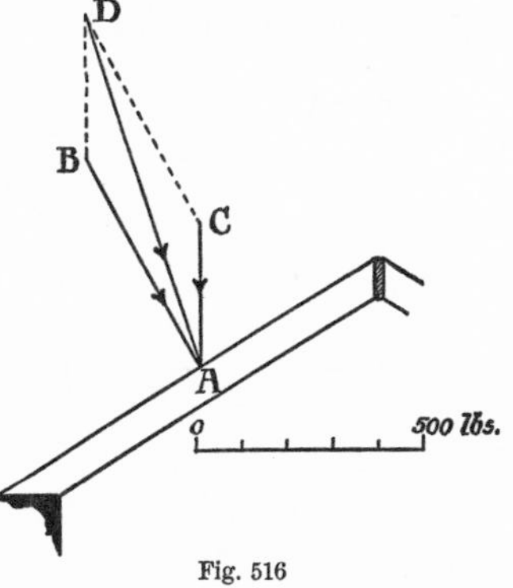

Fig. 516

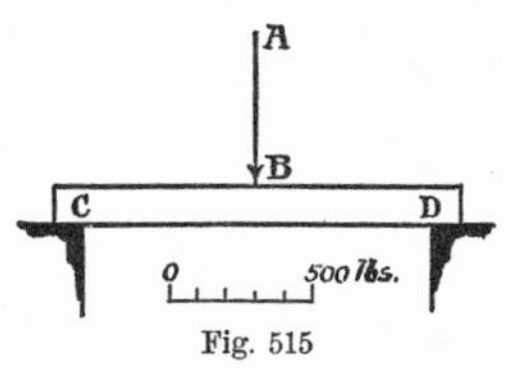

Fig. 515

1. The Parallelogram of Forces.—The parallelogram of forces represents a most important truth, which may be thus stated: *If two forces acting simultaneously at the same point be represented, in point of application, direction, and magnitude, by two adjacent sides of a parallelogram, their resultant will be represented by the diagonal of the parallelogram drawn from the point of application of the two forces.*

In fig. 516, two forces of 500 and 300 lbs. are represented by the lines B A and C A, the points of application being at A, the directions being indicated by the arrows, and the magnitudes by the length of the lines. The resultant of these two forces will be represented by the line D A, which is the diagonal of the parallelogram A B D C.

In other words, the force represented by the line A D will have exactly the same effect on the body at A as the two forces A B and A C.

2. THE TRIANGLE OF FORCES.—*If three forces be represented in magnitude and direction by the three sides of a triangle taken in order, they will balance each other if simultaneously applied at one point.*

Let *a*, *b*, and *c* (No. 1, fig. 517) be lines representing three forces in magnitude and direction. The lines taken in order form a triangle, and if applied simultaneously at one point (A, No. 2) will balance each other. The truth of this can be easily shown by finding the resultant of any two of the forces in No. 2, say *a* and *b*; it will be represented by the line *d*, which is equal and opposite to *c*.

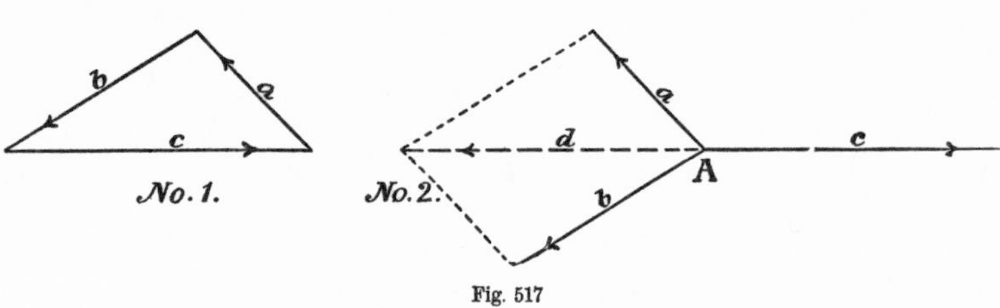

Fig. 517

Conversely—*Three forces which, when simultaneously applied at one point, balance each other, can be correctly represented in magnitude and direction by the three sides of a triangle taken in order.*

3. THE POLYGON OF FORCES.—*If any number of forces be represented in magnitude and direction by the sides of a polygon taken in order, they will balance each other if simultaneously applied at one point.*

Let *a*, *b*, *c*, *d*, and *e* (No. 1, fig. 518) be lines representing five forces in magnitude and direction. The lines taken in order form a polygon, and if applied simultaneously at one point (A, No. 2) will balance each other. The truth of this can be shown by finding the resultants of any two pairs of forces (say *d* and *e*, and *b* and *c*), and then finding the resultant of the two forces thus found; this final resultant will be represented by the line *f*, which is equal and opposite to the only remaining original force *a*.

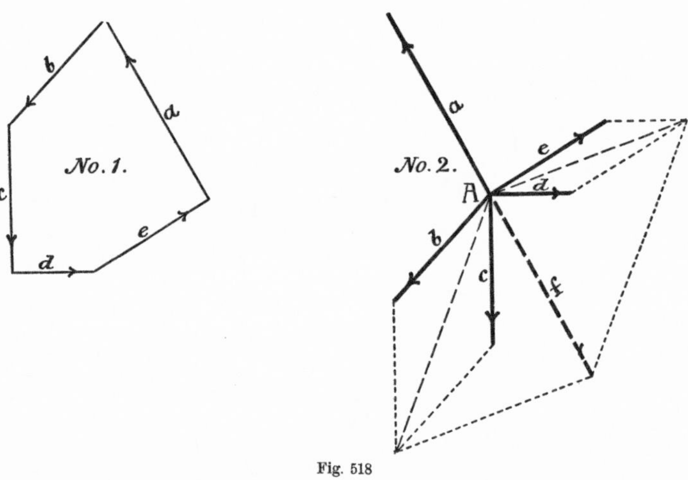

Fig. 518

Conversely—*Any number of forces which, when simultaneously applied at one point, balance each other, can be correctly represented in magnitude and direction by the sides of a polygon taken in order.*

It follows from this proposition that the resultant of any number of forces applied at one point and not in equilibrium, can be found by representing the forces by lines taken in order, to form so many sides of a polygon, and the line required to complete the polygon will represent the resultant force in magnitude and direction. Thus, if we have four forces represented by the lines *a*, *b*, *c*, and *d*, in fig. 518, No. 2, and it is required to find the resultant of these forces, draw the lines *a*, *b*, *c*, and *d*, in No. 1, in order of direction and parallel to the corresponding lines in No. 2, and equal to them each to each; the line *e*, which is required to complete the polygon, will represent the resultant force in magnitude and direction.

It is clear from what has been already written that, as two or more forces may have a single resultant, so a single force may be regarded as resolvable into two or more components. Thus, in fig. 516, the force D A is resolvable into the two component forces B A and C A, and also into an infinite number of other pairs of forces, as an infinite number of parallelograms can be drawn about the diagonal A D.

4. COMPOSITION OF PARALLEL FORCES.—*The magnitude of the resultant of any number of parallel forces is equal to the sum of the magnitudes of the forces, and its direction is the common centre of these forces provided they are symmetrically disposed.* Thus, in fig. 519, E is the resultant of the four parallel and symmetrically-disposed forces A, B, C, and D.

The Funicular Polygon.—If the forces are not symmetrically disposed, the resultant can be obtained by means of a polar diagram and funicular polygon. Let A B, B C, C D, D E, and E F (fig. 520, No. 1) be five parallel forces of the magnitudes shown. To find the resultant, set off the five forces in a straight line *af*, as shown in No. 3, parallel to the forces; take any pole *o*, and join *ao*, *bo*, &c. Produce the lines of force in No. 1, as shown by the dotted lines, dividing No. 2 into the spaces *a′*, *b′*, *c′*, *d′*, *e′*, and *f′*.

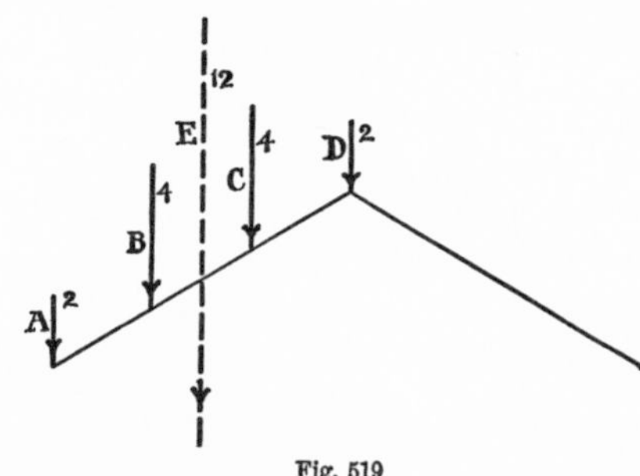

Fig. 519

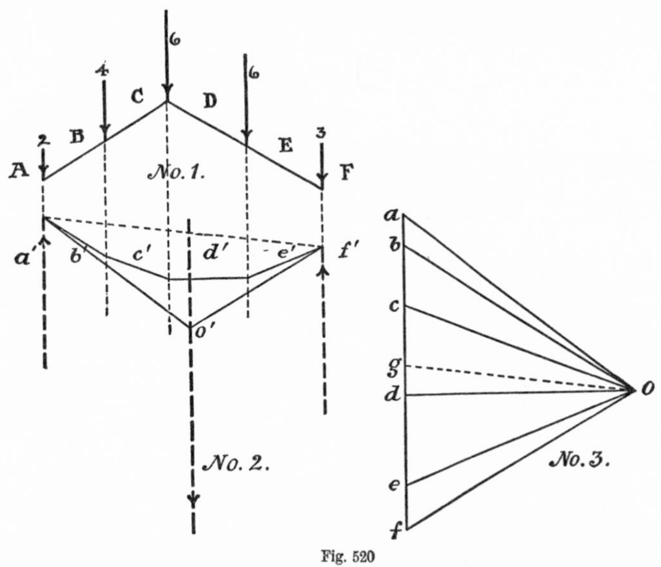

Fig. 520

From any point in *a′ b′* draw a line parallel to *bo* till it cuts the line *b′c′*; from the point of intersection draw another line parallel to *c o*, till it cuts *c′d′*; proceed in a similar manner for the other spaces, making the line in space *d′* parallel to *do*, and that in space *e′* parallel to *e o*. From the ends of the funicular polygon thus obtained draw lines parallel to *ao* and *fo* respectively, intersecting at *o*. The resultant will pass through the point of intersection, and its magnitude will be the sum of the five forces, that is to say, equal to *af* in No. 3.

If the ends of the funicular polygon are now closed as shown by the dotted line *a′f′*, and a line is drawn parallel to it from *o* in the polar diagram, it will cut *af* in *g*. If the inclined struts are supported at A and F, the supporting force or reaction at A will be represented by the line *ag*, and that at F by *fg*.

5. MOMENTS.—*The moment of a force about a given point is the product of the force and the perpendicular distance from the point to the line of action of the force.*

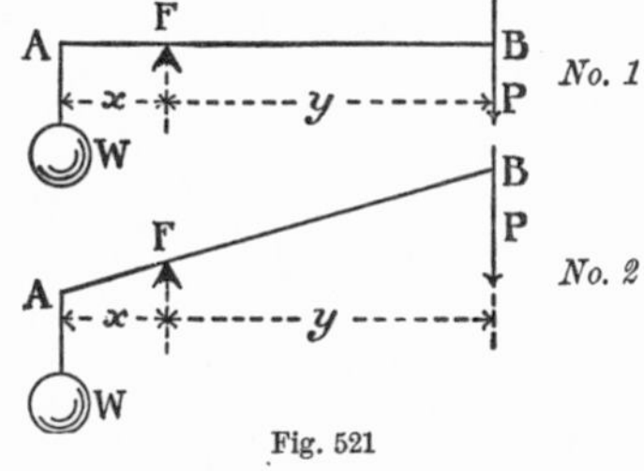

Fig. 521

The simplest example is that of the lever, fig. 521. The lever A B rests on the fulcrum F; a weight W is suspended at the end A, and a vertical force P is applied at the other end B. The moment of the force W about the point F is Wx, and the moment of the force P about the same point is Py. In No. 1 the lever is perpendicular to the lines of action of the two forces, and A F is therefore equal to x, and F B to y. In No. 2 the forces are oblique, and x is less than A F and y less than F B.

If these forces are in equilibrium, $W x = P y$, and $W = \frac{P y}{x}$, and $P = \frac{W x}{y}$.

Consequently, if $x = 1$ ft. and $y = 3$ ft., a weight of 60 lbs. at A will be balanced by a weight of 20 lbs. at B.

CHAPTER III

CANTILEVERS AND BEAMS

1. INTRODUCTORY

When a load is applied to a beam, the beam bends or tends to bend. This tendency is, however, resisted by the beam. We thus have two opposing forces, on the equilibrium of which the stability of the structure depends. At every point in the beam the moment of resistance must equal the bending moment; if the latter exceeds the former, the beam will break. The theory of the strength of beams is therefore based on the equation—

Greatest Bending Moment = Moment of Resistance.

The bending moment varies according to the length of the beam, the method of supporting or fixing it, and the nature and point or points of application of the load.

The moment of resistance varies according to the cross-section of the beam, and the coefficient of strength of the material of which the beam is made.

In the case of a beam supported at both ends, and loaded uniformly or in the middle, the tendency of the load is to cause the beam to bend downwards, as shown in fig. 522. A moment's consideration will show that in the lowest fibres of the beam a tensile stress is set up, and in the uppermost fibres a compressive stress, as indicated by the arrows. These stresses are most severe in the extreme fibres—that is to say, in those forming the soffit and top of the beam—and gradually diminish in intensity towards the central plane of the beam, until a layer of fibres is reached where the tensile and compressive stresses are alike zero.

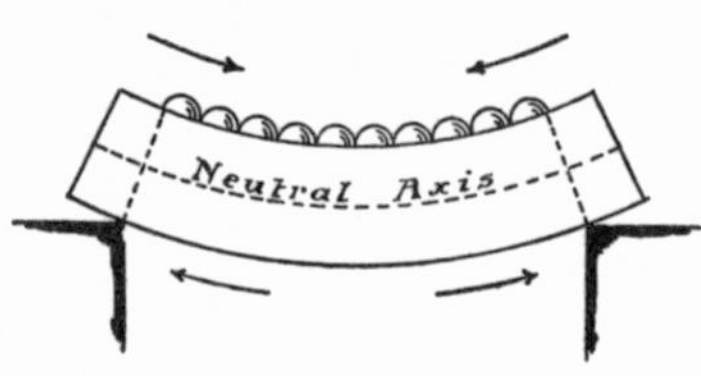

Fig. 522.—Tensile and Compressive Stresses in a Beam

This layer of fibres is known as the neutral plane or axis of the beam. In the case of materials whose resistance to tension is exactly equal to their resistance to compression, the neutral axis will be always at the centre of the beam, as shown in fig. 522; but as timber is much stronger in tension than in compression, the neutral axis of a wooden beam, loaded beyond its elastic limit, is gradually lowered with increasing loads, so that more and more fibres are brought under compression to counterbalance the greater tenacity of the material in the lower part of the beam. Barlow's experiments led him to the conclusion that, in a fir beam, loaded to rupture, the neutral axis is lowered to the plane at five-eighths of the depth of the beam. It is, however, the aim of the designer to proportion a beam so that it will not be loaded beyond its elastic limit, and in a rectangular beam so proportioned, the neutral axis is assumed to be along the central horizontal plane of the beam, and this position may be accepted as the basis of calculation for rectangular timber beams. The assumption may be thus stated: The neutral plane or axis of a horizontal beam is a horizontal plane passing through the centre of gravity of the cross-section. The centre of gravity of a rectangular section is at the point where the diagonals cross each other, and that of a triangle is found by drawing a line from the centre of one of the sides to the opposite angle, and taking a point on this line at a distance of two-thirds of its length from the angle and one-third from the side.

In addition to tensile and compressive stresses, shearing stresses are also set up under transverse stress. A beam may be considered as a series of horizontal fibrous layers, and these layers tend to slide upon each other, as shown at A in fig. 523. This tendency represents the horizontal shearing stress. Many beams loaded to rupture fail in this manner. A beam may also be considered as a series of vertical planes, as shown at B in fig. 523, with a tendency to slide upon each other in consequence of the vertical shearing stresses. The horizontal and vertical shearing stresses are equal to each other. As, however, the resistance of wood to shearing across the grain is far in excess of its resistance to shearing along the grain, the vertical shearing stresses need not be further considered.

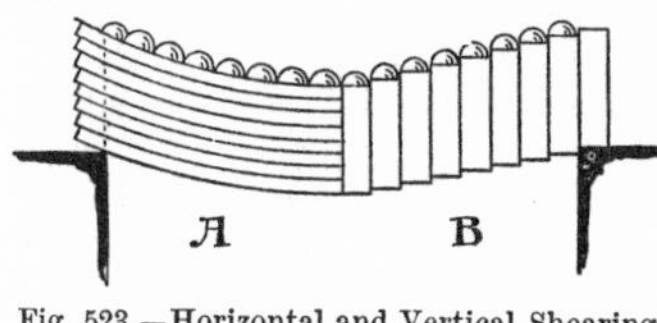

Fig. 523.—Horizontal and Vertical Shearing Stresses due to Transverse Stress

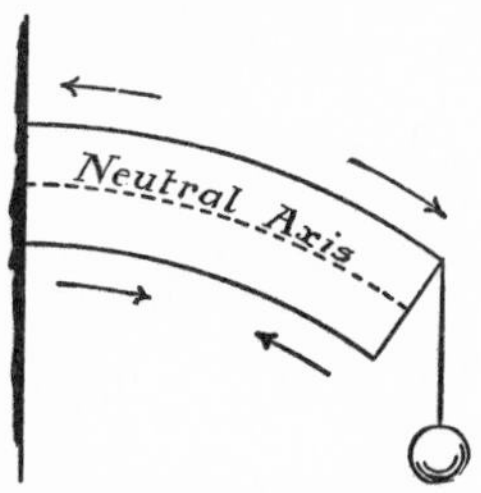

Fig. 524.—Tensile and Compressive Stresses in a Cantilever

A beam loaded uniformly or at one end, and fixed at the other end, is subject to similar stresses, but in these cases the upper fibres are in tension and the lower in compression, as shown in fig. 524.

2. BENDING MOMENTS AND SHEARING STRESSES

Beams may be supported or fixed in various ways:—1, Fixed at one end; 2, supported at both ends; 3, supported at two points at some distance from the ends; 4, fixed at both ends; 5, fixed at one end and supported at the other; and 6, supported or fixed at the ends, and supported at one or more intermediate points.

Loads may be distributed equally or unequally over the length of the beam, or concentrated at one or more points, or partly distributed and partly concentrated.

In the simplest cases, as will be hereafter shown, the calculations can be most easily made arithmetically by means of formulas; in others, however, the graphic method is less laborious. In a work of this kind it is impossible to pass in review every case which may occur in practice, but sufficient examples will be given to enable the student to deal with the more common forms of loads and supports.

(I) CANTILEVERS, OR BEAMS FIXED AT ONE END

Case 1.—*With the load concentrated at the free end.* The greatest bending moment in this case (fig. 525) will occur at A. Let M = bending moment at A, then $M = W l$. The bending moment at C, whose perpendicular distance x from B is $\frac{2}{3}l$, will be $\frac{2 W l}{3}$, and the moment at any other point can be found in a similar manner, or graphically as follows: Let $W = 1000$ lbs. (neglecting the weight of the beam) and $l = 96$ inches, then the moment at A will be $1000 \times 96 = 96{,}000$ inch-lbs. Draw the line ab equal to l, set off from a the line $a\,a$ perpendicular to $a\,b$, and mark on it (to any convenient scale) the moment at A in inch-lbs. Join $a_1\,b$. The bending moment at any intermediate point, as C, can be measured by scale at the corresponding position in the diagram, namely $c\,c_1$.

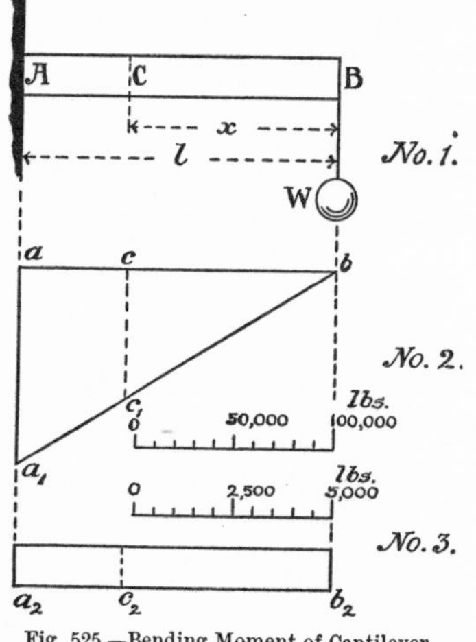

Fig. 525.—Bending Moment of Cantilever

The shearing force will be equal to W, and will be constant throughout the beam, as shown in No. 3. By this it is meant that at every vertical section of the beam there is a vertical shearing stress equal to W, that is to say—in this case—1000 lbs. The total horizontal shearing stress at any point is equal to the vertical shearing stress at that point, but the horizontal stress is not uniformly distributed throughout the cross section; it is greatest at the neutral axis, and can be represented graphically (for rectangular beams) by a parabola whose double ordinate is equal to the depth of the beam, and whose height or axis must be calculated from the formula—

$$\text{Maximum intensity of shearing stress} = \frac{3\,s}{2\,bd}$$

where s = vertical shearing stress, b = breadth of beam, and d = depth of beam. By the help of this formula the maximum intensity of horizontal shearing stress can be calculated from the shearing-stress diagram in any given case.

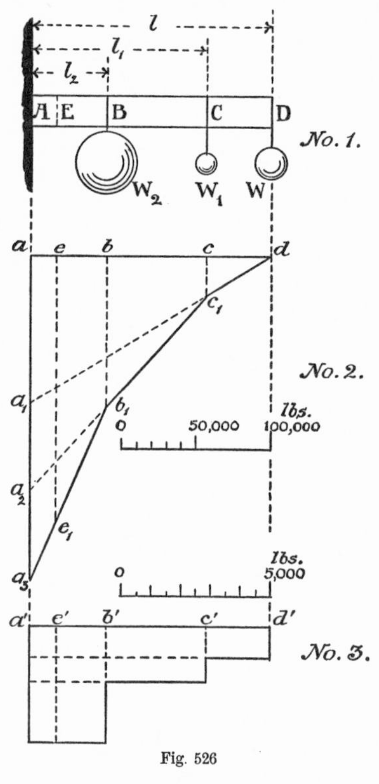

Fig. 526

Case 2.—*With two or more concentrated loads.* This is merely a variation of Case 1, and can be most easily understood by considering each load and length separately. Thus, the moment for the load w (fig. 526) is w l, that for w_1 is $w_1 l_1$, and that for w_2 is $w_2 l_2$. The stress diagram can be found by first plotting the moment for w, namely $a\,a_1\,d$; then that for w_1, namely $a_1\,a_2\,c_1$; and finally, that for w_2, namely $a_2\,a_3\,b_1$. The moment at any intermediate point, as E, can be measured by scale at the corresponding point in the moment diagram, namely ee_1. The stress diagram is calculated from the following data:—w = 1000 lbs., w_1 = 800 lbs., w_2 = 2000 lbs., l = 96 inches, l_1 = 70 inches, and l_2 = 30 inches.

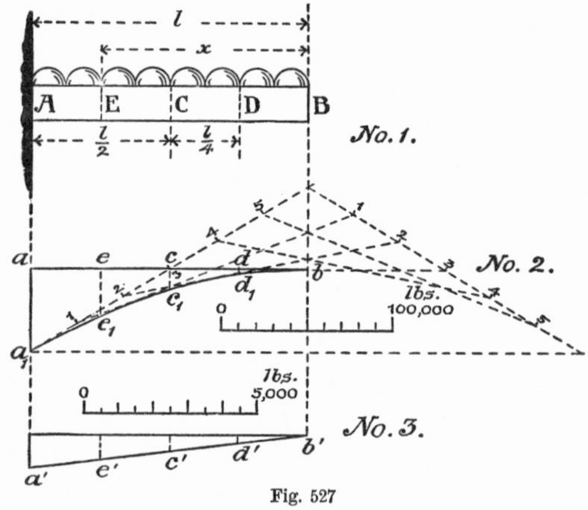

Fig. 527

The shearing stresses (No. 3) will be as follows:—

From C to D = W = 1000 lbs.
" B " C = W + W_1 = 1800 lbs.
" A " B = W + W_1 + W_2 = 3800 lbs.

Case 3.—*With the load uniformly distributed.* A load uniformly distributed may be considered as acting through its centre of gravity, namely, at the point C in fig. 527. If the load per unit of length is represented by w, the total load will be $w \times l$ = W. Then the bending moment at A will be $\frac{w\,l^2}{2}$, and as $w\,l$ = W, $\frac{w\,l^2}{2} = \frac{\text{W}\,l}{2}$. In other words, the maximum bending moment will be only one-half the moment which would be given by a load of equal weight concentrated at B.

To draw the diagram of moments, let w = 125 lbs. per foot, and l = 8 feet; then W = 125 × 8 = 1000 lbs., and the moment at A will be 1000 lbs. × $\frac{96 \text{ ins.}}{2}$ = 48,000 inch-lbs. Draw $a\,b$ = A B, and set off $a\,a_1$ = 48,000 to any convenient scale. The moment at any intermediate point, say C, will be found in a similar manner; the weight tending to bend the beam at this point will, however, be only the part of the load which lies between C and the free end B, in this case $\frac{\text{W}}{2}$. This part will act through its centre of gravity D, at a distance from C of $\frac{l}{4}$. The moment at C will therefore be $\frac{\text{W}}{2} \times \frac{l}{4} = \frac{\text{W}\,l}{8}$, or 12,000 inch-lbs. Plot this from c to c_1. The moments at other points can be found in a similar manner, and the graphic representation of the moments will be a parabolic curve, as shown at $a_1\,c_1\,b$.

The moment at any point may be expressed in general terms. Let a be the given point at a distance x from the free end, and let w = the weight per unit of length, then the moment at $a = w\,x \times \frac{x}{2} = \frac{w\,x^2}{2}$.

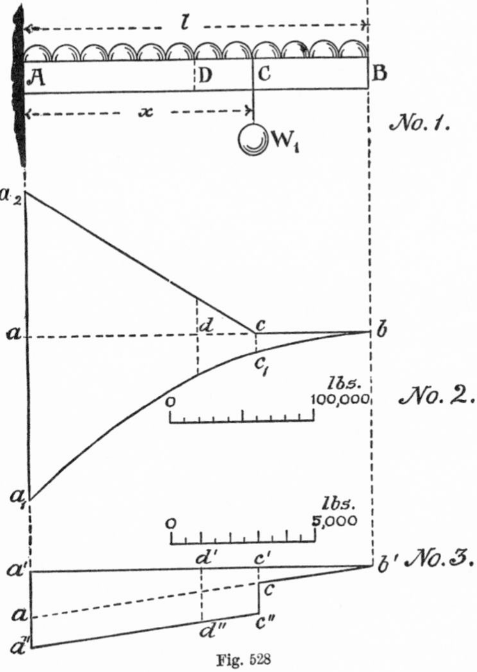

Fig. 528

The shearing stress (No. 3) at any point is equal to the load between the point and the free end, and consequently increases uniformly from the free to the fixed end; that is to say, from 0 to $w\,l$, or, in this case, from 0 to 1000 lbs.

Case 4.—*With a concentrated load and a load uniformly distributed.* This is merely a combination of Cases 1 and 3. Let w = the distributed load per unit of length, and W_1 the load concentrated at any point C. The moments for the distributed load will be found as in Case 3, and those for the concentrated load as in Case 1, and can be plotted together as shown in fig. 528, No. 2. In the illustration $w = 125$ lbs. per foot, $W_1 = 1000$ lbs., $l = 12$ feet, and $x = 8$ feet.

The shearing stresses will be a combination of those in Cases 1 and 3, and can be expressed graphically as in No. 3, fig. 528, where $a'\,a = w\,l =$ 1500 lbs., and $a\,a'' = W_1 = 1000$ lbs. The shearing stress at any point D can be measured by scale at $d'd''$.

(II.) BEAMS SUPPORTED AT BOTH ENDS

Case 5.—*Loaded in the centre.* The greatest bending moment will be at the centre of the beam. As action is equal and opposite to reaction, it follows that the upward reaction of the supports must be equal to the downward action of the load, and as the load is equidistant from the two supports, the reaction at each support will be equal to $\frac{W}{2}$. The moment of the upward reaction at A about the point B will be $\frac{W}{2} \times l = \frac{W\,l}{2}$, and the moment of the downward action of the load W about the same point B will be $W \times \frac{l}{2} = \frac{W\,l}{2}$. Consequently the reaction at A cannot raise the load W, and the half-beam A C may therefore be considered as a reversed cantilever, fixed at C and with an upward force applied at A. This reduces the problem to a simple duplication of Case 1, the load at A being $\frac{W}{2}$, and the length from A to C being $\frac{l}{2}$; the bending moment at C is therefore $\frac{W}{2} \times \frac{l}{2} = \frac{W\,l}{4}$.

Let W = 2000 lbs., and l = 16 feet; the bending moments of the half-beam A C may be plotted as at $a\,c\,c_1$ in No. 2, and those of the half-beam C B as at $c\,c_1 b$. The maximum bending moment will be the same as in Case 1, although both the load and span are doubled.

The shearing stresses (No. 3) will be equal to $\frac{W}{2}$ throughout each half-beam, but zero at the exact centre of the whole beam.

Case 6.—*With a concentrated load at any point on either side of the centre.* The greatest bending moment will be at the point where the load is applied, and the problem must be considered in a similar manner to Case 5. As, however, the load is not equidistant from the two supports, the reactions at the supports will not be equal. To find these reactions, let m and n be the two parts into which the length l is divided by the load. The greater portion of the load must obviously be borne by the support A, as this is nearer the load, and it is equally obvious that the two reactions will vary exactly as m and n;

that is to say, $R_a : R_b :: n : m$. And, as $R_a + R_b = W$, and $n + m = l$, $l : n :: W : R_a$, and $l . m :: W : R_b$. Consequently, if W, m, and n are known, the reactions can be determined by a simple rule-of-three sum. Thus, let W = 2000 lbs., m = 3 feet, and n = 7 feet, then l = 10 feet; and $10 : 7 :: 2000 : R_a$, $\therefore R_a = 1400$ lbs., and $10 : 3 :: 2000 : R_b$, $\therefore R_b = 600$ lbs.

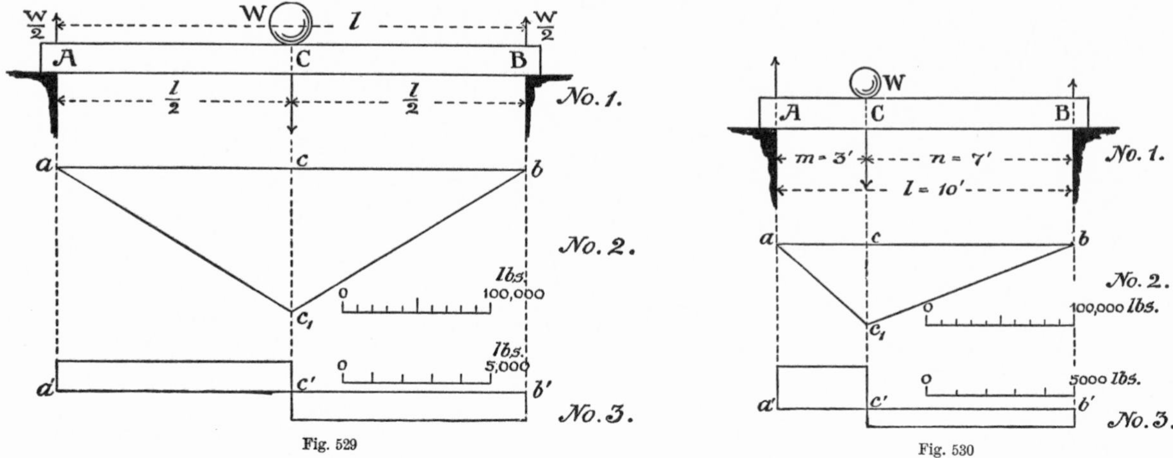

Fig. 529

Fig. 530

That this is correct can be easily proved by looking at the beam as a lever, as in Case 5; $R_a \times l = W \times n$, $\therefore R_a = \frac{Wn}{l} = \frac{2000 \times 7}{10} = 1400$ lbs. And $R_b \times l = W \times m$, and $R_b = \frac{Wm}{l} = \frac{2000 \times 3}{10} = 600$ lbs.

The reactions at the supports may therefore be conveniently expressed in general terms, thus—$R_a = \frac{Wn}{l}$, and $R_b = \frac{Wm}{l}$.

Having obtained the reactions, we now assume (as in Case 5) that the segment A C is a reversed cantilever fixed at C, and loaded at A with a load equal to the reaction at this support, namely $\frac{W}{l}$. The bending moment at C will therefore be $\frac{Wn}{l} \times m = \frac{Wmn}{l}$. Similarly, that for the segment C B will be $\frac{Wm}{l} \times n = \frac{Wmn}{l}$. In this example $\frac{Wmn}{l} = \frac{2000 \times 36 \times 84}{120} = 50{,}400$ inch-lbs. The moments are plotted $a\,c_1\,b$.

The shearing stress (No. 3) in each segment of the beam is equal to the reaction at the nearest point of support, and is uniform throughout the segment.

Case 7.—*With two or more concentrated loads.* This is simply a development of Case 6, and involves no new theory. It will be best, therefore, to work out the bending moments of a typical example. Let the points of application of the loads be as shown in fig. 531, and let W_1 = 1000 lbs., W_2 = 2000 lbs., and W_3 = 1500 lbs.

For the load W_1 the bending moment will be $\frac{1000 \text{ lbs.} \times 36 \text{ ins.} \times 144 \text{ ins.}}{180 \text{ ins.}} = 28{,}000$ inch-lbs.

For the load W_2, $M = \frac{2000 \text{ lbs.} \times 72 \text{ ins.} \times 108 \text{ ins.}}{180 \text{ ins.}} = 86{,}400$ inch-lbs.

For the load W_3, $M = \frac{1500 \text{ lbs.} \times 120 \text{ ins.} \times 60 \text{ ins.}}{180 \text{ ins.}} = 60{,}000$ inch-lbs

The first is plotted at $a\,c_1\,b$, the second at $a\,d_1\,b$, and the third at $a\,e_1\,b$.

The diagram of bending moments can now be completed by producing the lines $c\,c_1$, $d\,d_1$, and $e\,e_1$, and making each equal to the sum of the moments at the points C, D, and E. Thus, $c\,c_4$ must be made equal to $c\,c_1 + c\,c_2 + c\,c_3$, and so on. Join $a\,c_4$, $c_4\,d_4$, $d_4\,e_4$, and $e_4\,b$, and the diagram is complete.

The shearing stresses in the segments nearest the supports are equal to the reactions at these supports. $R_a = \left(1000 \times \frac{12}{15}\right) + \left(2000 \times \frac{9}{15}\right) + \left(1500 \times \frac{5}{15}\right) = 2500$ lbs., and

$R_b = \left(1500 \times \frac{10}{15}\right) + \left(2000 \times \frac{6}{15}\right) + \left(1000 \times \frac{3}{15}\right) = 2000$ lbs. The shearing stress in C D will be $R_a - W_1 = 2500 - 1000 = 1500$ lbs. That in D E will be $R_a - W_1 - W_2 = 2500 - 1000 - 2000 = -500$ lbs. The *minus* sign shows that the shearing stress in D E is in the opposite direction to that in C D. Similarly the shearing stress in E B may be said to be $= R_a - W_1 - W_2 - W_3 = 2500 - 1000 - 2000 - 1500 = -2000$ lbs. These figures are plotted in No. 3, fig. 531.

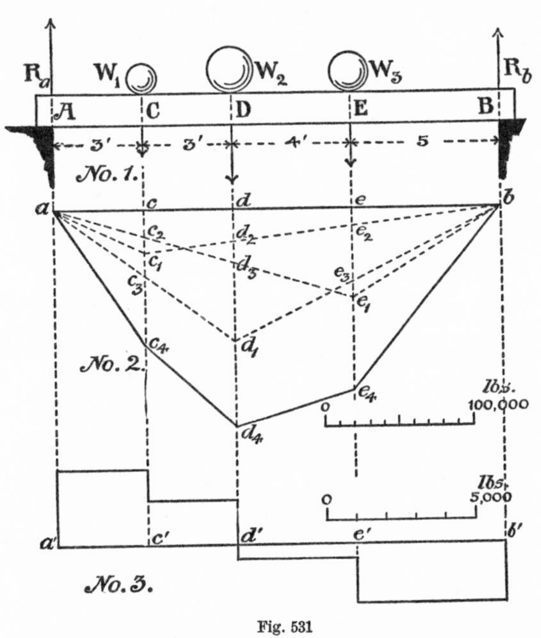

Fig. 531

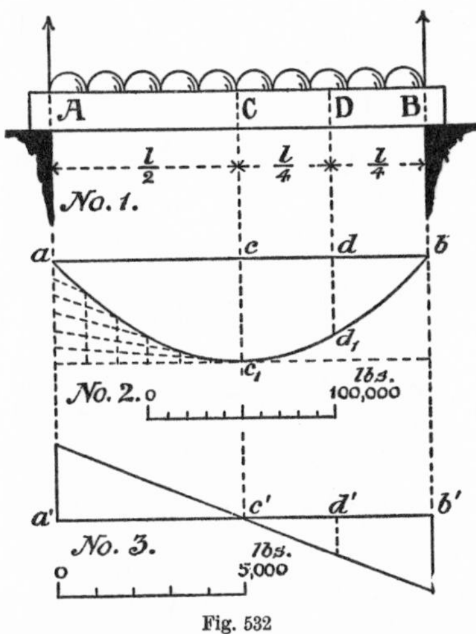

Fig. 532

Case 8.—*With load uniformly distributed.* The reaction at each support will be half the total load, that is to say, if w = the load per unit of length, the reaction at each support will be $\frac{wl}{2} = \frac{W}{2}$. As in Case 5, this problem can be best understood by considering the beam as two cantilevers fixed at the centre C (No. 1, fig. 532). The load on each half of the beam, as C B, can be considered as acting through its centre of gravity D, at a distance, therefore, of $\frac{l}{4}$ from C. The bending moment at C will be $\frac{wl}{2} \times \frac{l}{4} = \frac{wl^2}{8}$ or $\frac{Wl}{8}$, and this will be the greatest bending moment. The curve of moments, as in Case 3, will be represented by a parabola, which, if $w = 400$ lbs. per foot and $l = 10$ feet, will be as shown at $a\, c_1\, d_1\, b$ in No. 2.

The shearing stress at each support will equal the reaction at that support, and will diminish to zero at the centre of the beam. The stress at any point is found by subtracting the load between the point and the nearer support, from the reaction at the support, and can be shown graphically, as in diagram No. 3, fig. 532.

Case 9.—*With a concentrated load and a load uniformly distributed.* This is merely a combination of Cases 5 or 6 and 8. After grasping the method of procedure in Case 4, which is a combination of Cases 1 and 3, the student will have no difficulty in solving this problem.

Case 10.—*With load uniformly distributed over part of the beam.* If we imagine the distributed load ($wx = W$) to act through its centre of gravity E, we can obtain the reactions at each support, as in Case 6, namely $R_a = \frac{Wn}{l}$ or $\frac{wxn}{l}$, and $R_b = \frac{Wm}{l}$ or $\frac{wxm}{l}$.

If E coincides with the centre of the beam, the maximum bending moment will be at this point. In any other case the maximum moment will be at some point between E and the centre of the beam. This point may be obtained by dividing the length x, over which the load is distributed, into two parts proportional to m and n. Thus, in fig. 533; $m : n :: m_1 : n_1$. The proportional parts must, however, be in reverse positions; if m is to the left, m_1 must be to the right, and if n is to the right, n_1 must be to the left. The position of the maximum bending moment can therefore be found by simple arithmetic, thus, $l : m :: x : m_1$, or $l : n :: x : n_1$. In fig. 533 the maximum bending moment is therefore at F.[1].

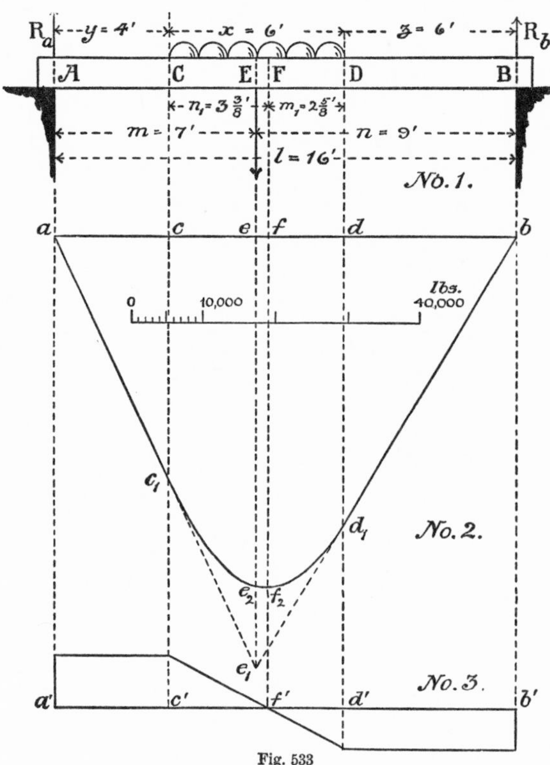

Fig. 533

The actual bending moments can be most easily solved graphically. First consider the load $w\,x$ to be a concentrated load acting through its centre of gravity E, and proceed as in Case 6. This will give the moment diagram $a\,e_1\,b$, which will be correct from a to c_1, and from b to d_1. Between c_1 and d_1 the line of moments will follow a parabolic curve. Join c_1 and d_1, and with $c_1\,d_1$ as double ordinate and e_1 as vertex describe the parabola $c_1\,e_2 f_2\,d_1$; this gives the curve of moments from C to D, and the maximum moment is shown to be at $f f_2$.

That this is correct can be proved mathematically by the use of the differential calculus, or in the following simple manner, which, as far as I know, is not mentioned in any text-book: Let $l = 16$ feet, $y = 4$ feet, $x = 6$ feet, $z = 6$ feet, and $w = 200$ lbs. per foot.

$$\therefore\ m = y + \frac{x}{2} = 4 + \frac{6}{2} = 7 \text{ feet, and } n = z + \frac{x}{2} = 6 + \frac{6}{2} = 9 \text{ feet.}$$

$$\therefore\ \text{as } l : n :: x : n_1,\ 16 : 9 :: 6 : n_1, \text{ and } n_1 = \frac{9 \times 6}{16} = 3\tfrac{3}{8} \text{ feet.}$$

The reaction at A $= \frac{w\,x\,n}{l} = \frac{200 \times 6 \times 9}{16} = 675$ lbs. We now assume, as in Case 5, that the part of the beam to the left of the position of the maximum bending moment, namely A F, is a reversed cantilever, fixed at F, and with an upward force R_a applied at A. The moment of this force will therefore be $R_a \times (y + n_1) = 675$ lbs. $\times$ (48 ins. + 40·5 ins.) = 59,737·5 inch-lbs. But against this moment must be set the opposite moment caused by the load between C and F. This load is $w \times n_1 = 200$ lbs. $\times\ 3\tfrac{3}{8}$ feet $= 675$ lbs. (that is to say, equal to the reaction at A), and it acts through its centre of gravity at a distance of $\frac{n_1}{2}$ from F. Therefore, the moment about the point F caused by the load on C F $= (w \times n_1) \times \frac{n_1}{2} = 675$ lbs. $\times \frac{40\tfrac{1}{2}\text{ ins.}}{2} = 13{,}668{\cdot}75$ inch-lbs.

The maximum bending moment at F will therefore be 59,737·5 – 13,668·75 = 46,068·75 inch-lbs. The accuracy of this is confirmed by applying the scale of inch-lbs. to the ordinate $f f_2$ in the diagram of bending moments, No. 2, fig. 533.

We are now in a position to state the maximum bending moment in general terms, the locus of this moment being known—

$$M = \left\{\frac{w\,x\,n}{l} \times (y + n_1)\right\} - \left(w\,n_1 \times \frac{n_1}{2}\right).$$

But we have seen that $\frac{w\,x\,n}{l} = R_a = w\,n_1$, therefore $M = R_a \times \left(y + \frac{n_1}{2}\right)$.

The shearing stress from A to C will be equal to the reaction at A, namely 675 lbs., and

[1] Two of the text-book formulas, altered to suit the notation adopted in fig. 533, are as follows:—

$$A\,F = y + x - z\left(\frac{2\,y + x}{2\,l}\right), \text{ and } A\,F = \frac{l^2 + y^2 - z^2}{2\,l}.$$

that from B to D will equal the reaction at B, namely 525 lbs. At F the shearing stress will be zero, and the diagram of shearing stresses will be as shown in No. 3, fig. 533.

(III.) COMBINED BEAMS AND CANTILEVERS SUPPORTED AT TWO POINTS

Case 11.—*With concentrated loads at the extremities of the cantilevers and centre of the span.* The cantilevers A C and B E are similar to Case 1. The bending moment at A (No. 1, fig. 534) produced by the load W_1 will be $W_1 \times l_1$, and that at B produced by the load W_2 will be $W_2 \times l_2$. These are shown graphically at $a\,a_1\,c$ and $b\,b_1\,e$. For convenience we assume that $l_1 = l_2$, and $w_1 = w_2$. The bending moments at A and B, produced by the two cantilevers, will produce an equal upward bending moment throughout the length l, as shown at $a\,a_2\,b_2\,b$. The central load W, as in Case 5, will produce a maximum downward bending moment at D equal to $\frac{Wl}{4}$; this is plotted from d_2 to d_1. Join $a_2\,d_1$ and $b_2\,d_1$, and the diagram of bending moments is complete.

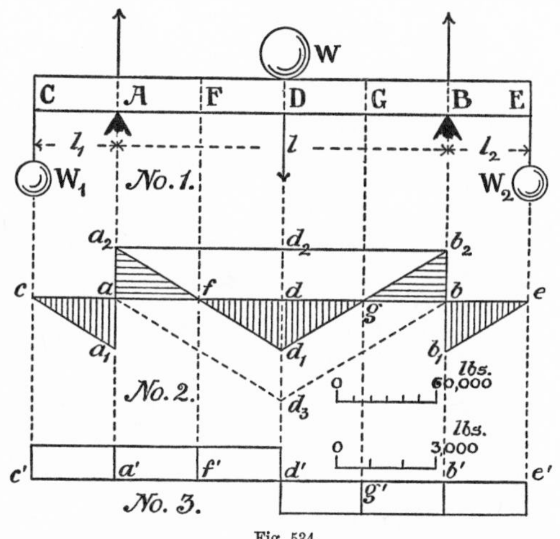

Fig. 534

It will be observed that much of the downward bending moment produced by the central load is neutralized by the upward bending moment produced by the cantilevers, the net bending moments being the shaded portions of the diagram, No. 2. The vertical shading represents the downward, and the horizontal shading the upward bending moments. At the points f and g in No. 2 (corresponding with F and G in No. 1) there is no bending moment. These points are known as the points of inflection or contraflexure.

In fig. 534, W = 2000 lbs., $W_1 = W_2 = 1000$ lbs., $l = 10$ feet, $l_1 = l_2 = 2$ feet 6 inches; l_1 and l_2 are made equal to each other, and each equal to $\frac{l}{4}$, and W_1 and W_2 are also equal to each other, and each equal to $\frac{W}{2}$, as this corresponds with what would occur if the cantilevers were omitted, and the beam were fixed at A and B, instead of being simply supported. The dotted lines $a\,d_3$ and $d_3\,b$ represent the diagram of bending moments in a beam A B, supported at the ends and centrally loaded with a weight W. It will be seen from this that the strength of the beam is doubled by fixing the ends, or by continuing the beam in the form of cantilevers equal in length to $\frac{l}{4}$ and loaded with weights at the free ends equal to $\frac{W}{2}$.

The shearing stresses are shown in No. 3, fig. 534. From C to A the shearing stress is uniform and equal to W_1, and from A to B it is uniform and equal to $\frac{W}{2}$. As in this case $W_1 = \frac{W}{2}$, the shearing stress is uniform throughout.

The reaction at A is equal to $W_1 + \frac{W}{2}$, and at B to $W_2 + \frac{W}{2}$.

Case 12.—*With the load uniformly distributed.* As in Case 3, the bending moment at A produced by the loaded cantilever A C (No. 1, fig. 535) is $w\,l_1 \times \frac{l_1}{2} = \frac{wl_1^2}{2}$. This produces an equal upward bending moment throughout the length l.

The maximum downward bending moment at E produced by the load $w\,l$, will, as in Case 8, be $\frac{wl}{2} \times \frac{l}{4} = \frac{wl^2}{8}$, and the bending moments throughout the remainder of the length l will form a parabolic curve, as shown at $a_2\,f\,e_1\,g\,b_2$ in No. 2.

As in Case 11, part of the downward bending moments caused by the load on A B are neutralized by the upward bending moment, so that the net downward bending moments are as shown by the vertically-shaded portions of the diagram. The horizontal shading represents the net upward bending moments. The points of contraflexure, where the bending moments are zero, are at f and g (F and G) at distances of $\frac{l}{2\sqrt{3}}$ from the centre of the beam, and the greatest bending moments are at the two supports A and B, namely, greatest downward moments immediately to the left of A and to the right of B, and greatest upward moments immediately to the right of A and to the left of B.

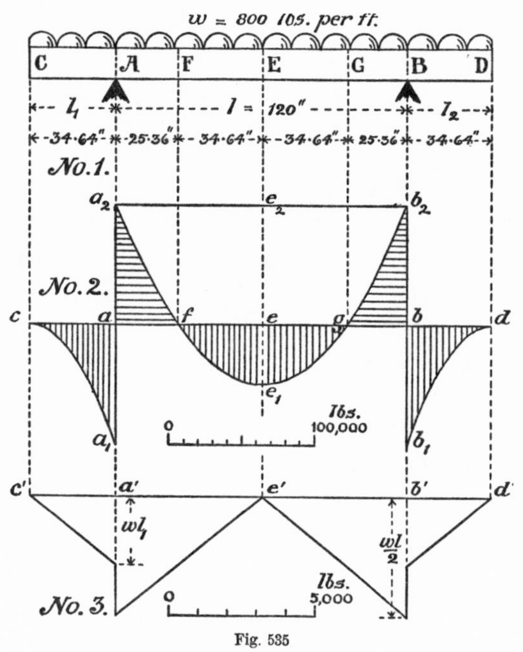

Fig. 535

In the illustrations the cantilevers are so proportioned as to give the same strength to the combination as if the ends of the beam were fixed and the cantilevers omitted.

The shearing stresses in the cantilevers increase from the free to the fixed end, directly as the weight, as in Case 3, and those in the beam increase from the centre to the support as in Case 8.

(IV.) BEAMS FIXED AT BOTH ENDS

Case 13.—*With concentrated load at centre.* The fixing of the beam at both ends reduces the maximum bending moment exactly one-half. No reduction, however, is effected in the shearing stress. This will be understood by reference to Case 11, fig. 534. The portions of the beam between the points of contraflexure and the supports, namely F A and G B, are by the fixing converted into cantilevers, and the real span of the beam is therefore reduced from A B to F G. As F G = $\frac{\text{A B}}{2}$, and as the strength of a beam varies inversely as the length, the strength of a beam is doubled by fixing the ends.

Case 14.—*With the load uniformly distributed.* The fixing of the ends reduces the maximum bending moment one-third, and the beam is therefore one-and-a-half times as strong as if the ends were merely supported. The shearing stress is not affected. See Case 12, fig. 535. It will be observed that the greatest bending moment is at each end, namely $a\,a_2$ and $b\,b_2$, and this bending moment is equal to $\frac{w\,l^2}{12}$, whereas, if the ends had been merely supported, the greatest bending moment would have been at the centre of the beam, and equal to $\frac{w\,l^2}{8}$, as explained in Case 8. The bending moment at the centre is $\frac{w\,l^2}{24}$.

(V.) BEAMS FIXED AT ONE END AND SUPPORTED AT THE OTHER

Case 15.—*With a concentrated load at the centre.* This case seldom occurs in carpentry, and need not be considered in detail. The reaction at the supported end, B (No. 1, fig. 536), will be only $\frac{5}{16}$ W. The bending moment at C will be $\frac{5\text{ W}}{16} \times \frac{l}{2} = \frac{5\text{ W}\,l}{32}$. The portion A D of the beam will act as a cantilever fixed at A and loaded at D with a load = $\frac{11\text{ W}}{16}$; the bending moment at A will therefore be $\frac{11\text{ W}}{16} \times \frac{3\,l}{11} = \frac{3\text{ W}\,l}{16}$. If, again, we assume D C to be fixed at C, and loaded at D with $\frac{11\text{ W}}{16}$, the moment at C will be $\frac{11\text{ W}}{16} \times \frac{5\,l}{22} = \frac{5\text{ W}\,l}{32}$, which is exactly equal to the moment produced at C by the segment C B.

The moment diagram (where W = 2000 lbs. and l = 120 inches) will therefore be as drawn in No. 2, fig. 536, $a\,a_1$ being the maximum bending moment (upward) and $c_1\,c_2$ being the maximum downward bending moment ($= \frac{5}{6}\,a\,a_1$). The point of contraflexure is at D. By fixing one end the maximum bending moment is reduced from $\frac{W\,l}{4}$ (Case 5) to $\frac{3\,W\,l}{16}$, a reduction of one-fourth, and the strength of a beam fixed at one end and supported at the other is therefore one-and-a-third times as strong as a similar beam supported at both ends.

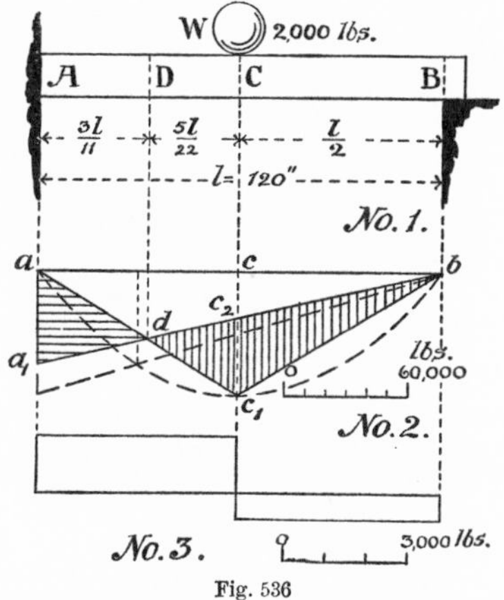

Fig. 536

The shearing stresses are equal to the reactions at B and D. From C to B the shearing stress is $\frac{5}{16}$ W, and from C to D and D to A it is $\frac{11}{16}$ W.

Case 16.—*With the load uniformly distributed.* In this case the strength of the beam is not increased by fixing one end (see Case 8), but the maximum bending moment is shifted from the centre of the beam to the fixed end. The moment at the fixed end $= \frac{w\,l^2}{8} = \frac{W\,l}{8}$, and at the centre $= \frac{W\,l}{16}$. The point of contraflexure is approximately at a distance of $\frac{l}{4}$ from the fixed end. The dotted oblique line and parabola in fig. 536 (No. 2) show the bending moments for a distributed load of 4000 lbs. = 2 W. The shearing stress is $\frac{3}{8}\,w\,l$ at the supported end and $\frac{5}{8}\,w\,l$ at the fixed end.

(VI.) CONTINUOUS BEAMS

Beams are said to be "continuous" when they are supported at each end and at one or more intermediate places. The calculations involved are too intricate for this work, and it will therefore be sufficient to state some of the general results which can be obtained.

Case 17.—*Continuous beam of two equal spans with the load uniformly distributed.*—Part of the beam on each side of the support B (No. 1, fig. 537) acts as a cantilever, and thus shortens the effective span of the beam and increases its strength.

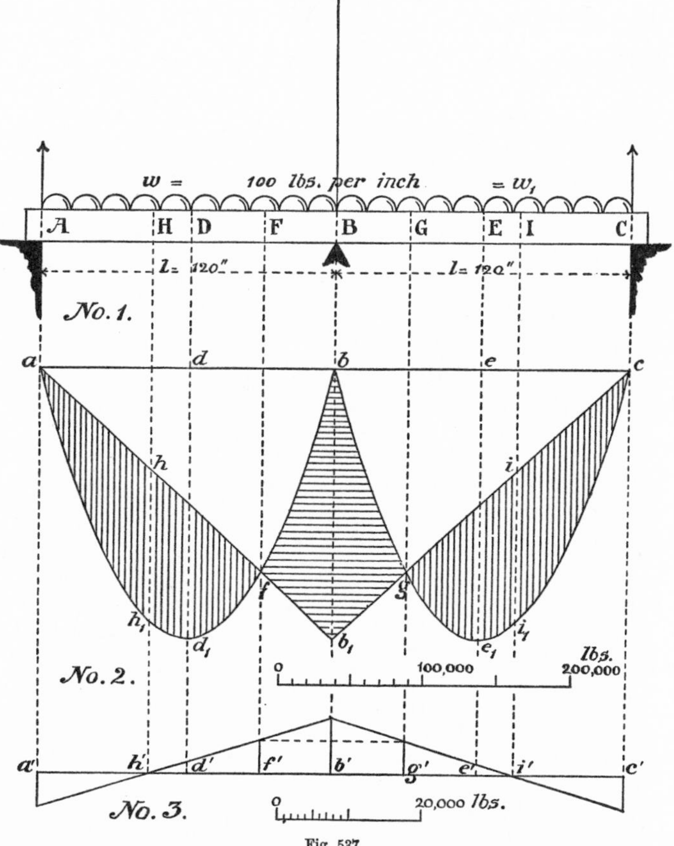

Fig. 537

The reactions at the supports are as follows:—$R_a = \frac{3}{8}\,w\,l$, $R_b = \frac{5}{4}\,w\,l$, and $R_c = \frac{3}{8}\,w\,l$.

By the theorem of three moments the moments at any three consecutive supports of a continuous beam can be obtained. This theorem—adopting the notification of fig. 537—is as follows, for beams of uniform section, with the load on each span uniformly distributed over that span:—$4\,l\,M_A + 8\,(l + l_1)\,M_B + 4\,l_1\,M_C = w\,l^3 + w_1\,l_1^3$. We know that at the extreme supports, A and C, the bending moment is zero, as in Case 8; consequently

$4 l M_A = 0$, and $4 l_1 M_C = 0$. The equation thus becomes, for a beam of two spans:— $8 (l + l_1) M_B = w l^3 + w_1 l_1^3$. If the two spans are equal, and both spans are uniformly loaded, as in fig. 537, we get:—

$$8 \times 2 l M_B = 2 w l^3, \text{ and } M_B = \frac{2 w l^3}{16 l} = \frac{w l^2}{8}.$$

If $w = 100$ lbs. per inch, and $l = 120$ inches, then $M_B = \frac{100 \times 120^2}{8} = 18{,}000$ inch-lbs.

This is the upward-bending moment at the support B, and is the maximum moment in the beam. It is shown in the moment diagram at $a b_1 c$.

The downward-bending moments on the two spans must now be calculated as if each half of the beams were supported at the ends and uniformly loaded, exactly as in Case 8. The maximum moment in each segment will be at the centre of the span, and will be $\frac{w l^2}{8}$. These must be plotted downwards from $a b$ and $b c$ as shown at $d d_1$ and $e e_1$, and the parabolic curves of the moments completed by the lines $a d_1 b$ and $b e_1 c$. The downward moments in the areas $a f b$ and $b g c$ are counteracted by corresponding upward moments, and the actual moments are therefore the shaded portions of the diagram, the horizontal shading representing upward moments and the vertical representing downward.

The points of contraflexure are at f and g (F and G), and the actual maximum moments in the outer portions of the beam are midway between A and F and between G and C, that is to say, at H and I, shown at $h h_1$ and $i i_1$ in the moment diagram. These, however, are of little importance in the case of beams of uniform section, as they are much less than the bending moment over the central support.

We can now see that the whole beam may be considered as made up of a number of segments or parts. A F may be considered as a beam supported at A and F, and uniformly loaded with w lbs. per unit, and giving therefore a reaction at $F = \frac{w \times AF}{2}$. F B is a cantilever uniformly loaded with w lbs. per unit, and also having a concentrated load at F equal to the reaction at this point caused by the load on A F. B G is similar to F B, and G C to A F.

The shearing stresses can now be determined as follows:—

$$S_A = \frac{w \times AF}{2};\ S_B = \frac{w \times AF}{2} + (w \times FB);\ S_C = \frac{w \times GC}{2}.$$

These are plotted in the shearing-stress diagram, No. 3, fig. 540. The reactions at the outer supports are equal to the shearing stresses, each to each; the shearing stresses close to the right and left of the central support are together equal to the reaction at that support.

It is unnecessary to enter into the detailed calculations of uniformly-loaded beams of three or more spans. The reactions and bending moments at the supports are as follows, where l = length of one span, and w = load per unit of length:—

REACTIONS AND BENDING MOMENTS AT SUPPORTS OF UNIFORMLY-LOADED CONTINUOUS BEAMS

	Three Equal Spans.		Four Equal Spans.		Five Equal Spans.	
1st Support, A	$R_A = \frac{4}{10} w l$	$M_A = 0$	$R_A = \frac{11}{28} w l$	$M_A = 0$	$R_A = \frac{15}{38} w l$	$M_A = 0$
2nd „ B	$R_B = \frac{11}{10} w l$	$M_B = -\frac{w l^2}{10}$	$R_B = \frac{8}{7} w l$	$M_B = -\frac{3 w l^2}{28}$	$R_B = \frac{43}{38} w l$	$M_B = -\frac{2}{19} w l^2$
3rd „ C	$R_C = \frac{11}{10} w l$	$M_C = -\frac{w l^2}{10}$	$R_C = \frac{13}{14} w l$	$M_C = -\frac{w l^2}{14}$	$R_C = \frac{37}{38} w l$	$M_C = -\frac{3}{38} w l^2$
4th „ D	$R_D = \frac{4}{10} w l$	$M_D = 0$	$R_D = \frac{8}{7} w l$	$M_D = -\frac{3 w l^2}{28}$	$R_D = \frac{37}{38} w l$	$M_D = -\frac{3}{38} w l^2$
5th „ E	...	...	$R_E = \frac{11}{28} w l$	$M_E = 0$	$R_E = \frac{43}{38} w l$	$M_E = -\frac{2}{19} w l^2$
6th „ F	..	...	...	...	$R_F = \frac{15}{38} w l$	$M_F = 0$

The reactions at the outer supports are equal to the shearing stresses, each to each; the reaction at any intermediate support is equal to the shearing stresses close to the right or left of that support taken together.

3. MOMENTS OF RESISTANCE

Within the limit of elasticity of the beam the moment of resistance is obviously equal to the bending moment, but if gradually increasing loads are applied, as in the case of a test-specimen, the bending moment eventually exceeds the moment of resistance, and the beam ultimately breaks. What is therefore required is a knowledge of the greatest moment of resistance which a beam can with safety be called upon to develop.

In the first portion of this chapter it was shown that, in a loaded beam supported at the ends, the upper part is in compression and the lower in tension, and that along the line of junction of these parts of the beam there is a plane, known as the neutral axis, where there is neither tension nor compression. If we carefully consider fig. 522 (page 304), we shall see that the greatest compressive stress will be in the uppermost layer of fibres, as this layer is most shortened by the bending of the beam, and that the stress gradually diminishes as the layers approach the neutral axis. Similarly, the greatest tensile stress will be in the lowest layer of fibres, as this is most elongated by the bending of the beam, and from this layer to the neutral axis the stress will gradually diminish. Let fig. 538 represent an enlarged cross-section of the beam; then the tensile and compressive stresses can be represented graphically by the triangles C E D and A E B, the greatest compressive stress being in the layer A B, and the greatest tensile stress in C D; from these extreme fibres the stresses gradually diminish towards the centre of gravity E of the cross-section, where the stress is zero.

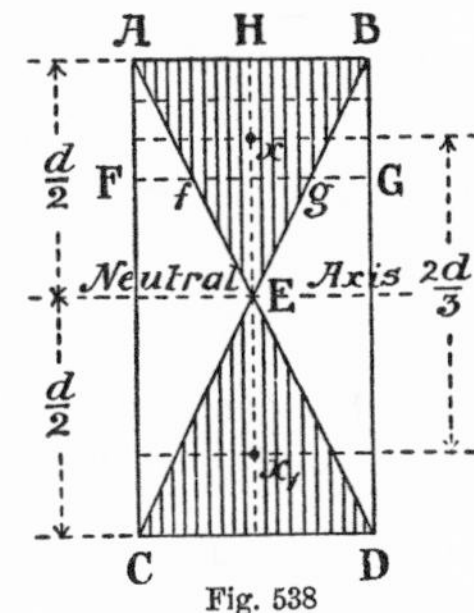

Fig. 538

It must be clearly understood that in the stress diagram, shown by the shading, the stress is uniform per unit, say, 1000 lbs. per square inch; thus, if the beam is 6 inches broad and 12 inches deep, the stress on the uppermost inch of the depth will be $5\frac{1}{2} \times 1 \times 1000 = 5500$ lbs., and this is distributed over an area of 6 square inches, so that the average stress on the uppermost inch of the beam is $\frac{5500}{6} = 916{\cdot}6$ lbs. per square inch. But at F G, which represents the centre of an inch-thick layer at one-fourth the depth of the beam, the stress fg is only 3 inches × 1 inch × 1000 lbs. = 3000 lbs., which, being distributed over 6 square inches as before, gives an average stress of only 500 lbs. per square inch.

The stresses in the diagram being equal in intensity, can be considered as acting through the centres of gravity of their respective areas, that is to say, the compressive stresses in the triangle A E B may be considered as acting at the centre of gravity x of this triangle, the centre of gravity being at a point distant from H one-third of the length H E. Similarly, the tensile stresses may be considered as acting at x_1. These stresses, therefore, may be looked upon as forming a couple with a length of arm $x\,x_1 = \frac{2}{3}d$.

But the stresses A E B are represented by a triangle whose area is $\frac{b\,d}{4}$. The moment of the couple is therefore $\frac{b\,d}{4} \times \frac{2}{3}d = \frac{b\,d^2}{6}$. If the intensity of the stress per unit (generally per square inch) is represented by f, we get

$$\text{Moment of Resistance} = \frac{f\,b\,d^2}{6}.$$

This formula is, however, only applicable to rectangular beams, with two of the sides placed vertically. It is not correct for beams of triangular, hexagonal, circular, or other section. A formula for general application must be more general in its terms, and must take into consideration what is known as the "moment of inertia" of the different sections.

The term "moment of inertia" originated in a wrong conception of the properties of matter. It has, however, been retained as a very convenient one, although the conceptions under which it originated have long ago vanished. "The meaning of the term as at present used, in relation to a solid body, is as follows:—*The moment of inertia of a body about a given axis is the limit of the sum of the products of the weight of each of the elementary particles that make up the body, by the squares of their distances from the given axis.*"[1]

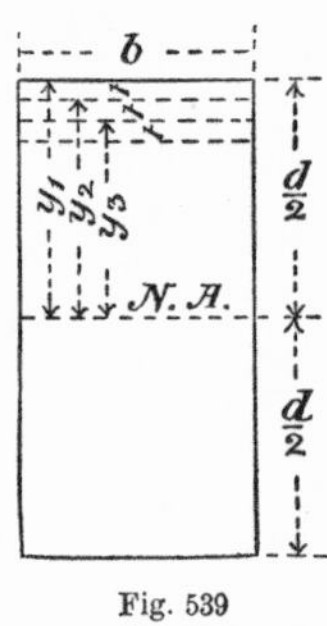

Fig. 539

The cross-section of the beam shown in fig. 539 may be considered as made up of a series of extremely thin horizontal layers. Let t = the thickness of each layer; y_1, y_2, y_3, &c., the distances of the several layers from the neutral axis; and I = the moment of inertia. Then $I = bty_1^2 + bty_2^2 + bty_3^2 + \ldots\ldots$, or $I = b(ty_1^2 + ty_2^2 + ty_3^2 + \ldots\ldots)$

This is conveniently expressed, thus: $I = b\Sigma ty^2$.

Now let f denote the intensity of stress at the distance y from the neutral axis, and f_1 the intensity at the distance y_1, then (according to fig. 538)

$$f_1 : f :: y_1 : y, \text{ and } f_1 = \frac{fy_1}{y}.$$

The moment of resistance of a layer is its direct resistance multiplied by its distance y_1 from the neutral axis, and its direct resistance is its breadth b × its thickness t × its greatest resistance to stress within the elastic limit of the material, say f_1. Therefore,

$$M_R = bt \times y_1 \times f_1 = bty_1f_1.$$

But $f_1 = \frac{fy_1}{y}$, therefore $M_R = bty_1 \times \frac{fy_1}{y} = \frac{f}{y}bty_1^2$.

Proceeding in a similar manner for succeeding layers, we obtain the formula

$$M_R = \frac{f}{y}b(ty_1^2 + ty_2^2 + ty_3^2 + \ldots\ldots), \text{ or } M_R = \frac{f}{y}b\Sigma ty^2.$$

But $b\Sigma ty^2$ = Moment of Inertia = I. Therefore, $M_R = \frac{fI}{y}$, and $f = \frac{M_R y}{I}$.

The following table gives the moments of inertia of beams of various sections:—

No.	Description and Illustration.	Area.	Moment of Inertia = I.	Distance of Neutral Axis from the extreme Fibres = y.
1.	Rectangle .... (N, A, b, d)	bd	$\frac{bd^3}{12}$	$\frac{d}{2}$
2.	Regular hexagon .... (N, A, s)	$\frac{3s^2\sqrt{3}}{2}$	$\cdot541\,s^4$	$\cdot866\,s$
3.	Regular octagon .... (N, A, r)	$2\cdot828r^2$	$\cdot638\,r^4$	$\cdot924r$
4.	Circle .... (N, A, r)	πr^2	$\frac{\pi r^4}{4}$	......

As an example, let us find the moment of resistance of a rectangular pitch-pine beam, 6 inches broad and 12 inches deep, the coefficient of transverse rupture (*i.e.* the value

[1] Professor Lanza, *Applied Mechanics.*

of f) being 6000 lbs. The formula is $M_R = \frac{fI}{y}$. According to the table (No. 1), $I = \frac{bd^3}{12}$, and $y = \frac{d}{2}$; therefore, $M_R = \frac{fbd^3}{12} \div \frac{d}{2} = \frac{fbd^2}{6}$, and $M_R = \frac{6000 \times 6 \times 12^2}{6} = 864{,}000$ inch-lbs.

4. FORMULAS

Having now found both the greatest bending moments and the moments of resistance of beams, we can place the results together and educe formulas for the more simple methods of loading. Only rectangular sections need be considered here. Throughout the formulas, f = coefficient of transverse rupture ascertained by actual test, w = breaking weight, b = breadth of beam in inches, d = depth of beam in inches, and l = length of clear span in inches.

I. *Beams fixed at one end and loaded at the other.* The greatest bending moment (see Case 1, page 305) is wl, and the moment of resistance of a rectangular beam is $\frac{fbd^2}{6}$. Therefore,

$$wl = \frac{fbd^2}{6}, \text{ and } w = \frac{fbd^2}{6l} \quad (1)$$

II. *Beams fixed at one end and loaded uniformly.* The greatest bending moment (Case 3) is $\frac{wl}{2}$. Therefore,

$$\frac{wl}{2} = \frac{fbd^2}{6}, \text{ and } w = \frac{fbd^2}{3l} \quad (2)$$

III. *Beams supported at both ends and loaded in the centre.* (See Case 5.)

$$\frac{wl}{4} = \frac{fbd^2}{6}, \text{ and } w = \frac{2fbd^2}{3l} \quad (3)$$

IV. *Beams supported at both ends and loaded at a point at a distance m from one support and n from the other.* (See Case 6.)

$$\frac{wmn}{l} = \frac{fbd^2}{6}, \text{ and } w = \frac{fbd^2l}{6mn} \quad (4)$$

V. *Beams supported at both ends and loaded uniformly.* (See Case 8.)

$$\frac{wl}{8} = \frac{fbd^2}{6}, \text{ and } w = \frac{4fbd^2}{3l} \quad (5)$$

VI. *Beams fixed at both ends and loaded in the centre.* (See Case 13.)

$$\frac{wl}{8} = \frac{fbd^2}{6}, \text{ and } w = \frac{4fbd^2}{3l} \quad (6)$$

VII. *Beams fixed at both ends and loaded uniformly.* (See Case 14.)

$$\frac{wl}{12} = \frac{fbd^2}{6}, \text{ and } w = \frac{2fbd^2}{l} \quad (7)$$

VIII. *Beams fixed at one end and supported at the other, and loaded in the centre.* (See Case 15.)

$$\frac{3wl}{16} = \frac{fbd^2}{6}, \text{ and } w = \frac{8fbd^2}{9l} \quad (8)$$

IX. *Beams fixed at one end and supported at the other, and uniformly loaded.* (See Case 16.)

$$\frac{wl}{8} = \frac{fbd^2}{6}, \text{ and } w = \frac{4fbd^2}{3l} \quad (9)$$

X. *Continuous beams of two equal spans, supported at the ends and centre, and loaded uniformly.* (See Case 17.) Let l = the length of each span, and W = $w\,l$ = the total load on each span.

$$\frac{\text{W}\,l}{8} = \frac{f\,b\,d^2}{6}, \text{ and W} = \frac{4f\,b\,d^2}{3\,l} \qquad (10)$$

For beams of three equal spans, $\frac{\text{W}\,l}{10} = \frac{f\,b\,d^2}{6}$, and W $= \frac{5f\,b\,d^2}{3\,l}$ .... (11)

For beams of four equal spans, $\frac{3\,\text{W}\,l}{28} = \frac{f\,b\,d^2}{6}$, and W $= \frac{14f\,b\,d^2}{9\,l}$ .... (12)

All these formulas can of course be transposed. Thus, No. 1 will give the following equations:—

$$\text{W} = \frac{f\,b\,d^2}{6\,l} \ldots(1). \quad f = \frac{6\,\text{W}\,l}{b\,d^2} \ldots(2). \quad b = \frac{6\,\text{W}\,l}{f\,d^2} \ldots(3).$$

$$d^2 = \frac{6\,\text{W}\,l}{f\,b} \ldots(4). \quad l = \frac{f\,b\,d^2}{6\,\text{W}} \ldots(5). \quad 6 = \frac{f\,b\,d^2}{\text{W}\,l} \ldots(6).$$

The value of f, which is known as the coefficient or modulus of rupture, varies for woods of different species and also for wood of the same species according to its seasoning and quality. It is ascertained by experiment, and will be fully considered in Part II of this Section.

It ought to be pointed out that the increase of strength due to fixing the ends of beams (Formulas 6 to 9) is not generally taken into consideration in carpentry, as there is great difficulty in making the ends of timber beams absolutely rigid. If, for example, the ends are built into walls, they may decay or shrink and thus become merely supported and not fixed; similarly, if they are bolted down, the stress on the bolts may be relieved by the shrinkage of the wood, or the fibres of the wood may yield under the pressure of the washers or nuts. In ordinary cases, therefore, it is better to assume that the ends are merely supported. In the case of cantilevers (Nos. 1 and 2), the fixing of the ends is of course a necessary feature, and must be carefully effected.

Formulas for shearing stresses and deflection will be given in Part II of this Section, Chapter IV.

CHAPTER IV

STRESSES IN FRAMED STRUCTURES

The term "truss" is frequently applied to a framed structure designed to support a load. In carpentry the principal trusses are those constituting the heavy framing of roofs, but there are also trussed beams, trussed partitions and walls, bridge-trusses, and the trusses for the centring of arches and for other purposes. It would lead us too far to consider in detail all these kinds of timber framing, but the main principles on which calculations of strength are based must be considered in order that the student may understand the parts which the different members of a truss are called upon to perform, and may be able to design the members and their joints in a rational manner. As roof-trusses are those most commonly required, they will be most fully considered.

Three kinds of stresses may occur in trusses, namely, tension, compression, and transverse stress or bending. Members in tension are known as ties or braces, and members in compression may be described as struts, although other names are applied to some of these members according to the position they occupy in the truss. Thus, king-posts and queen-posts are really ties, being subjected to tensile stresses, and principal rafters, being under compression, are struts. In addition to tensile and compressive stresses, some of the members of trusses are often called upon to bear a transverse stress; a tie-beam carrying a floor or ceiling is a case in point.

A truss must consist of at least three pieces, arranged to form a triangle, but it will be useful to consider first the stresses set up by a load applied at the angle formed by two bars only.

Case 1.—*Two struts* (fig. 540). Let A B and B C be the two inclined members of the frame, supported at A and C, and loaded at B with a weight represented in magnitude, direction, and point of application by the line B D.

From D draw D E parallel to B C and meeting A B in E, and from D also draw D F parallel to A B and meeting B C in F. The force B D has been resolved into two components, namely B F acting along the line B C, and B E acting along the line B A. In other words, the load at B causes in B C a stress equal to BF, and in B A a stress equal to B E, these stresses being measured by the same scale of lbs. or cwts. or tons adopted in laying down the force-line B D. It is obvious that these stresses are compressive, and that, if the joint at B were hinged or pivoted, the abutments at A and C would be overturned if they were not of sufficient weight or strength to resist the thrusts.

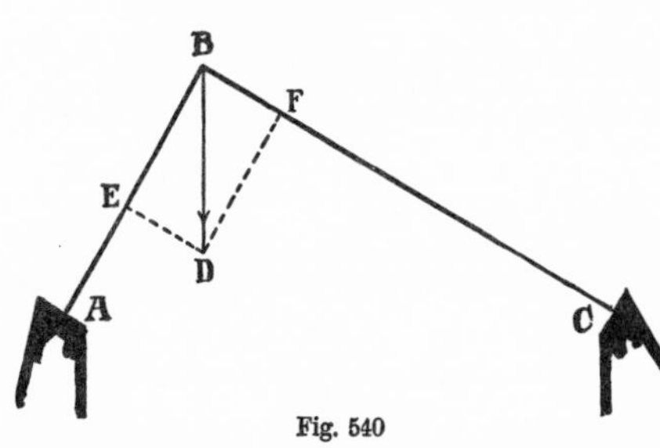

Fig. 540

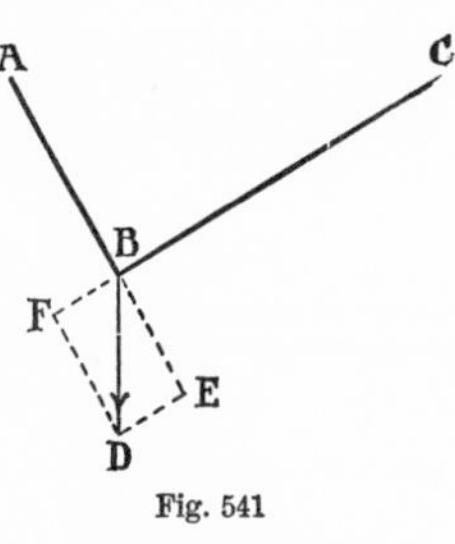

Fig. 541

Case 2.—*Two ties* (fig. 541). Let A B and B C be the two inclined members of a frame, suspended at A and C, and loaded at B with a weight represented in magnitude, direction, and point of application by the line B D.

From D draw D E parallel to B C and meeting A B produced in E, and from D also draw D F parallel to A B and meeting C B produced in F. The force B D has been resolved into two components B E and B F, of which B E is equal to the pull or tensile stress in the tie A B, and B F is equal to that in B C.

Case 3.—*A strut and a tie* (fig. 542). Let A B and B C be the two inclined members of the frame, suspended at A and supported at C, and loaded at B with a weight represented by the line B D.

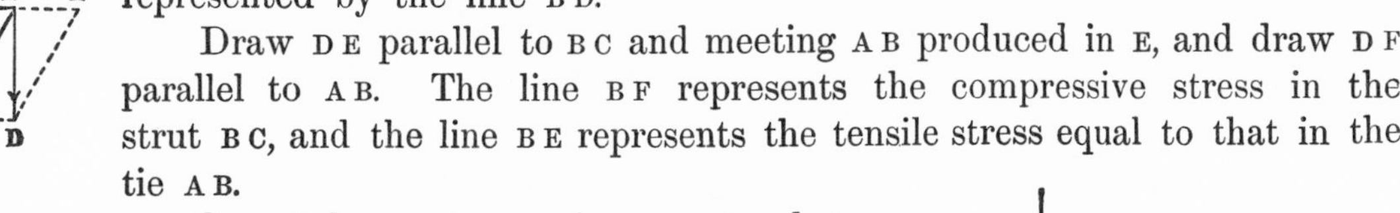

Fig. 542

Draw D E parallel to B C and meeting A B produced in E, and draw D F parallel to A B. The line B F represents the compressive stress in the strut B C, and the line B E represents the tensile stress equal to that in the tie A B.

Case 2 does not occur in carpentry, but Case 1 represents the common form of weaving-shed roof, and Case 3 occurs in the jib and stay rope of a crane and also in framed brackets supporting galleries and balconies.

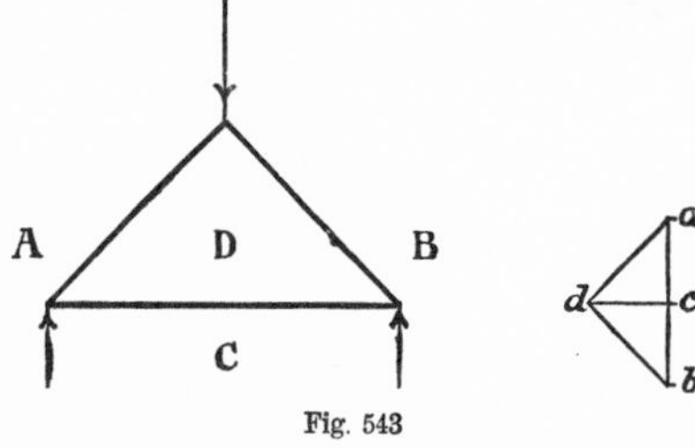

Fig. 543

Case 4.—*Triangular truss with concentrated load* (fig. 543). In this and subsequent illustrations Bow's ingenious system of notation is adopted, which consists in allotting a capital letter to each *space* in the truss, and to each of the surrounding spaces, while in the stress diagram small letters are placed at the *angles.* Each member of the truss is known by the letters in the spaces on each side of it; thus, in fig. 543 the tie-beam is C D, the left-hand rafter A D, and the right-hand rafter B D. It will be found that this system of notation simplifies the stress diagrams in a very happy manner.

Let D be a triangular truss loaded at the apex with a weight represented in magnitude, direction, and point of application by the vertical line at the apex, and supported at the two extremities of the tie-beam. As the load is applied centrally between the supports, the reaction at each support will be $\frac{W}{2}$. We thus have three known external forces, all acting vertically, and therefore parallel to each other. These forces must first be plotted. The load at the apex lies between the spaces A and B, and is therefore plotted as *a b* in the stress diagram. The left-hand supporting force lies between the spaces A and C, and is shown by the line *a c*, the right-hand supporting force being shown by the line *b c*. As the three external forces are parallel to each other, the triangle of external forces *a b c* is a straight line.

In finding the stresses in a truss it is best to adopt a certain order whenever possible. We will begin at the left-hand support, and proceed in the direction taken by the hands of a clock. The supporting force has been plotted at *c a* parallel to C A; from *a* draw *a d* parallel to A D, and from *c* draw *c d* parallel to C D. This gives the triangle of forces *c a d* at the left-hand support, *c a* representing the supporting force, *a d* the stress in A D, and *d c* the stress in D C; *c a* is, we know, an upward force, and the others follow in order around the triangle. The force *a d* is from the bar A D *towards* the joint at the left-hand support, and A D is therefore in compression; the force *d c* is *from* the joint towards the bar D C, and D C is therefore in tension.

To complete the stress diagram, from *d* draw *d b* parallel to D B; if the diagram is correctly drawn, *d b* will meet *a b* in *b*. The triangle of forces for the apex is *a b d*, and that for the right-hand support is *b c d*. The direction of these forces is not shown in the diagram, as they would render it less clear; thus, at the right-hand support *b c* is an upward force, *c d* a horizontal force in the direction *from* the joint at this support, and *d b* is an inclined downward force *towards* the joint.

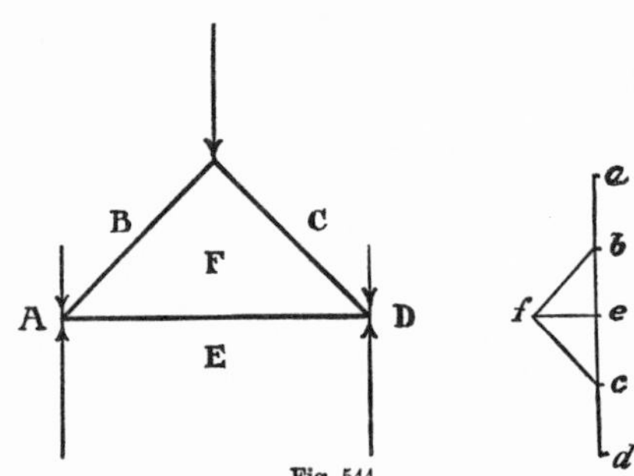

Fig. 544

Case 5.—*Triangular truss with distributed load* (fig. 544). In this case it is obvious that the rafters A F and B F are subjected to transverse stresses in addition to direct tension and compression. Complicated stresses of this kind cannot be accurately calculated, and it ought to be the aim of the designer to avoid such transverse stresses by concentrating the loads at the joints of the truss. In a simple triangular frame like that in fig. 544, which would only be used for very small spans, the concentration of the load is not very material, and the transverse stresses will not now be taken into consideration.

If we represent the total distributed vertical load by the letter W, the load on B F will be $\frac{W}{2}$, and this may be regarded as made up of two equal vertical loads (each $= \frac{W}{4}$) applied at the apex and left-hand support respectively. The load on C F may be regarded in a similar manner. We thus have three external downward forces due to the load, namely $\frac{W}{4}$ at the left-hand support, $\frac{W}{4} + \frac{W}{4} = \frac{W}{2}$ at the apex, and $\frac{W}{4}$ at the right-hand support. As the apex of the truss is vertically over the centre of the span, the supporting forces opposed to the load are each $\frac{W}{2}$. We thus have five known external forces, and as these are all vertical, the polygon of external forces is a straight line *a b*, *b c*, *c d*, *d e*, *e a*.

The forces at the left-hand support already known are *a b* and *a e*. From *b* draw *b f* parallel to B F, and from *e* draw *e f* parallel to E F. The stress in B F is represented by the line *b f*, and that in E F by *e f*. The directions of the stresses follow in order around the polygon of forces; A B we know to be a downward force, and this gives the clue to the directions of the other forces. The stress *b f* is seen to be in an oblique downward direction, and shows that the stress in B F is *towards* the joint at A; B F is therefore in compression.

The stress fe is horizontal from left to right, and shows that the stress in FE is from the joint at A; FE is therefore in tension.

To find the remaining stress at the apex draw cf parallel to CF; $bcfb$ is the stress diagram for the joint at the apex, and $cdefc$ for the right-hand support.

Case 6.—*Unequal-sided truss with concentrated load* (fig. 545). In this case the left-hand supporting force will be $\frac{Wn}{l}$, and the right-hand supporting force will be $\frac{Wm}{l}$, as explained in the preceding chapter. The supporting forces being known, the three external forces can be plotted at $abca$, and the procedure is then the same as in Case 4. It is clear from the stress diagram that the reactions can also be found graphically; draw ab equal

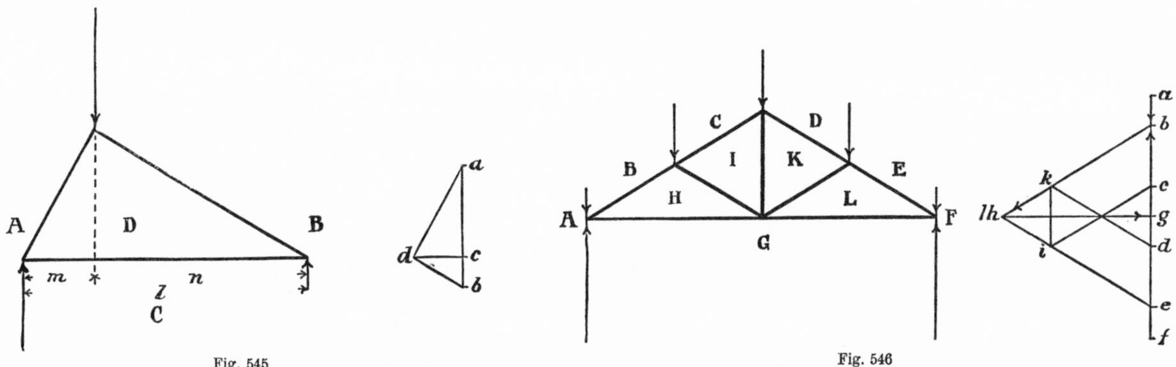

Fig. 545

Fig. 546

and parallel to the load on the apex of the truss; from a draw ad parallel to AD, from b draw bd parallel to BD, and from d draw dc parallel to DC. The line ab is divided into two parts, ac and cb, which give the reactions at AC and CB respectively.

Case 7.—*King-post truss with uniform load concentrated at the joints* (fig. 546). The load is supposed to be uniformly distributed over the common rafters, and to be concentrated at the supports, centres of rafters, and apex by means of pole-plates, purlins, and ridge-piece.

If the total load be W, the load on the left-hand rafter will be $\frac{W}{2}$, and one-half of this (*i.e.* $\frac{W}{4}$) will be concentrated by the purlin at the centre of the rafter, while the other half will be shared equally between the left-hand supporting joint and the apex. The load on the other rafter will be similar. We thus have $AB = \frac{W}{8}$, $BC = \frac{W}{4}$, $CD = \frac{W}{8} + \frac{W}{8} = \frac{W}{4}$, $DE = \frac{W}{4}$, and $EF = \frac{W}{8}$. Each supporting force will be $\frac{W}{2}$. These seven external forces are all vertical, and their polygon is therefore a straight line, namely $abcdefga$.

Commencing at the left-hand joint, we have the known external forces ga, ab; from b draw bh parallel to BH, and from g draw gh parallel to GH. The stress diagram for the left-hand joint is therefore $gabhg$, ga being an upward force, ab a downward force, bh an oblique downward force *towards* the joint (BH is therefore in compression), and hg a horizontal force *from* the joint (HG is therefore in tension).

At the joint BCIH we have four forces, two of which are known, namely, bc and bh; we can therefore find the other two. From h draw hi parallel to HI, and from c draw ci parallel to CI. The polygon of forces is $bcihb$, bc being a downward force, ci an oblique downward force *towards* the joint (CI is therefore in compression), ih an oblique upward force *towards* the joint (IH is therefore in compression), and hb, which had previously been determined.

The stress diagram for the apex is completed by drawing dk parallel to DK, and ik parallel to IK. The polygon of forces is $cdkic$, cd being a downward force, dk an upward force *towards* the joint (DK is therefore in compression), and ik a downward force *from* the joint (IK is therefore in tension). The remaining portions of the stress diagram do not call for explanation.

Case 8.—*Queen-post truss with uniform load concentrated at the joints* (fig. 547). Let W = the total load. Then, the load being concentrated at equidistant points, we have the following downward external forces:—AB = $\frac{W}{12}$, BC = $\frac{W}{6}$, CD = $\frac{W}{6}$, DE = $\frac{W}{6}$, EF = $\frac{W}{6}$, FG = $\frac{W}{6}$, and GH = $\frac{W}{12}$. The upward external forces are AI = $\frac{W}{2}$, and IH = $\frac{W}{2}$. The polygon of external forces is therefore the straight line *a b c d e f g h i a*.

Commencing at the left-hand support, we have the two known forces *i a* and *a b*. From *b* draw *b k* parallel to BK, and from *i* draw *i k* parallel to IK. The polygon of forces at this joint is *i a b k i*; BK is in compression, and IK in tension. The polygon of forces at the next joint is *b c l k b*; CL, LK, and KB are all in compression, the forces being towards the joint.

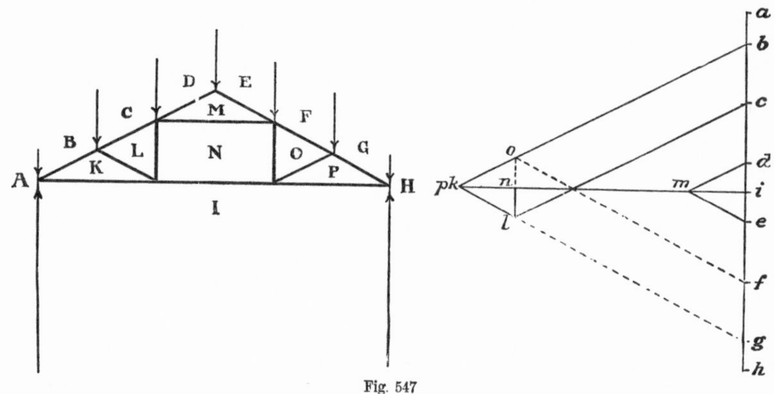

Fig. 547

At the next joint there are five forces, only two of which are known, namely, CD and CL. Before we can determine the stress diagram for this joint, it is necessary to ascertain another of the five forces. This can be done by drawing the triangle of forces for the apex, namely *d e m d*. This gives us the stress in DM; this member is in compression. The stresses at the joint CDMNL can now be ascertained by drawing *m n* parallel to MN, and *n l* parallel to NL. The polygon of forces at this joint is therefore *c d m n l c*, from which it will be gathered that MN is in compression, and NL in tension. To avoid confusion, the remainder of the stress diagram is shown by dotted lines.

Case 9.—*King- and queen-post truss with uniform load concentrated at the joints* (fig. 548). This example of a more complicated truss is introduced, as it involves a different order

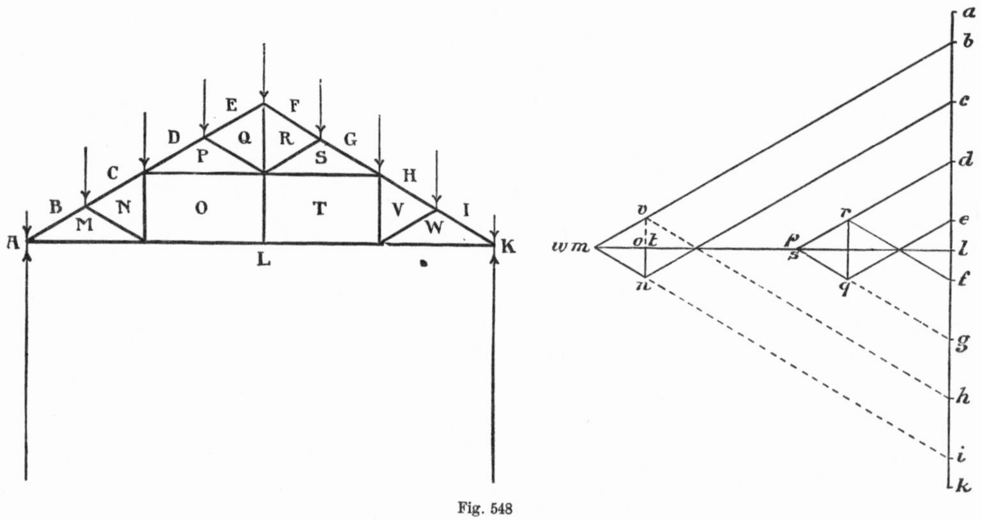

Fig. 548

of procedure; it contains a member which, if the tie-beam were not loaded, would be theoretically superfluous. The load at each support is $\frac{W}{16}$, and at each of the other joints $\frac{W}{8}$; each supporting force is $\frac{W}{2}$. The stress diagram for the joint at the left-hand support is *labml*, and that for the first joint in the rafter is *bcnmb*. It will be observed that there are three unknown forces at each of the remaining joints in this rafter, including the ridge, and the stress diagrams cannot therefore be drawn. At the tie-beam joint LMNO two forces are known, and the remaining two can be found by drawing the stress diagram *lm*, *mn*, *no*, and *ol*, or *lmnol*. This gives the stress in NO, and the polygon of forces for the middle joint of the rafter can now be drawn, namely *cdponc*. For the next joint the polygon of forces is *deqpd*, and that for the apex is *efrqe*.

If the diagram is completed as shown by the dotted lines, it will be found that *o* and *t* coincide, and consequently there is no stress in OT. Trusses of this design were erected over the City Hall at Glasgow, as illustrated in the section on Carpentry, but the tie-beams were used to support the ceiling, and the member OT would therefore be of service in order to prevent the sagging of the middle half of the tie-beam. If the stress diagram is carefully considered, it will be found that the tension-members of the truss are the tie-beam, the two queen-posts, and the king-post; all the other members, including PO and ST, are in compression.

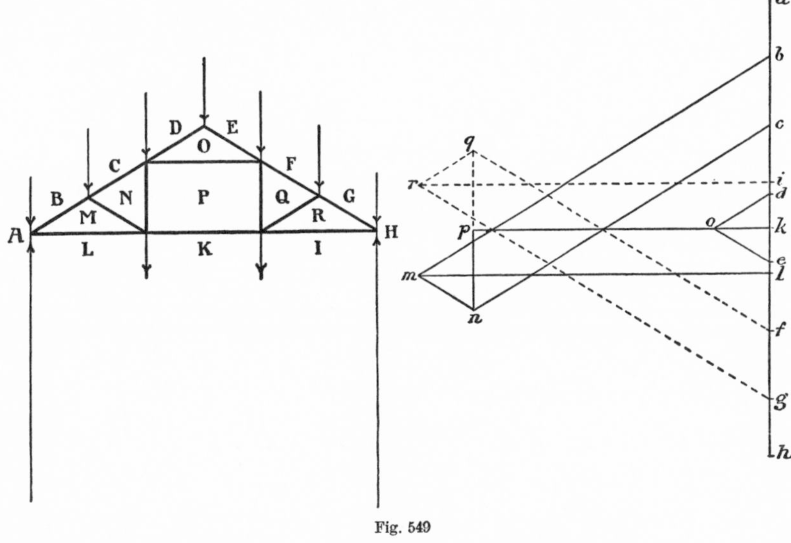

Fig. 549

Case 10.—*Queen-post truss with uniform load concentrated at the joints, and with the tie-beam uniformly loaded* (fig. 549). The loading of the tie-beam introduces a new element, as briefly indicated in the last example. Let the total load on the truss = W, of which $\frac{1}{4}$ is borne uniformly and directly by the tie-beam, and the remainder concentrated by the purlins, pole-plate, and ridge-piece at the upper joints of the truss. The external load ($\frac{3}{4}$ W) will be concentrated as follows:—AB $= \frac{W}{16}$, BC $= \frac{W}{8}$, CD $= \frac{W}{8}$, DE $= \frac{W}{8}$, and similarly for the other half of the truss. The load on the tie-beam $\left(\frac{W}{4}\right)$ may be regarded (for the purpose of ascertaining the stresses at the joints) as concentrated at the two supports and at the feet of the queen-posts, so that we shall have LK $= \frac{W}{12}$, KI $= \frac{W}{12}$, and at each support we shall have an additional load of $\frac{W}{24}$ due to the loading of the tie-beam. The downward force AB is therefore $\frac{W}{16} + \frac{W}{24} = \frac{5}{48}$W. IH is the same. The upward reaction at each support is of course $\frac{W}{2}$. The eleven external forces being now known, the stress diagram can be commenced.

The polygon of external forces is the straight line *abcdefghikla*. The side *hi* represents the upward force HI, *ik* is the downward force IK, *kl* the downward force KL, and *la* the upward force LA. Beginning at the left-hand support, we have the polygon of forces *labml*. For the first purlin joint we obtain *bcnmb*. For the foot of the left-hand queen-post we have three known forces, *kl*, *lm*, and *mn*, and the other two are found by drawing *np* and *pk* parallel to NP and PK respectively, completing the polygon *klmnpk*. For the second purlin joint we obtain the polygon *cdopnc*, and for the apex *deod*.

If this stress diagram is compared with that in fig. 547, the differences caused by the load on the tie-beam will be plainly seen. The cross-bending stress due to the distributed load on the tie-beam must in practice be taken into consideration, the tie-beam being a continuous girder supported at the ends and at two intermediate points.

Case 11.—*Truss with tie-beam and collar-beam* (fig. 550). This case presents no difficulty, and is introduced merely for purposes of comparison with Case 12. Note that the tie-beam H G is in tension, and the collar H K in compression.

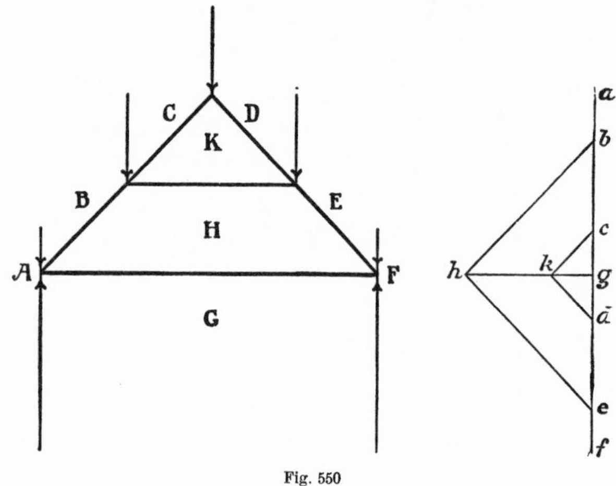

Fig. 550

Case 12.—*Collar-beam truss* (fig. 551). Trusses of this kind are often used for small spans, each pair of common rafters having its own collar. In all graphic determinations of stresses in trusses it is assumed that the members are free to move at the joints, as if the joints were formed simply with pins or pivots. If we consider fig. 551 in the light of this assumption, it is obvious that the truss would spread as shown in No. 2, unless a tie were inserted, as at H G in fig. 550, or the horizontal thrust were counteracted by suitable abutments. In practice the rafter is in one piece from end to end, and its rigidity determines in a large measure the stability of the roof. Ignoring this for the moment, we will assume that abutments are to be provided.

As the direction of the supporting force at A and the magnitude of the stress in B G are unknown, we cannot begin the stress diagram at the left-hand support. Let us commence

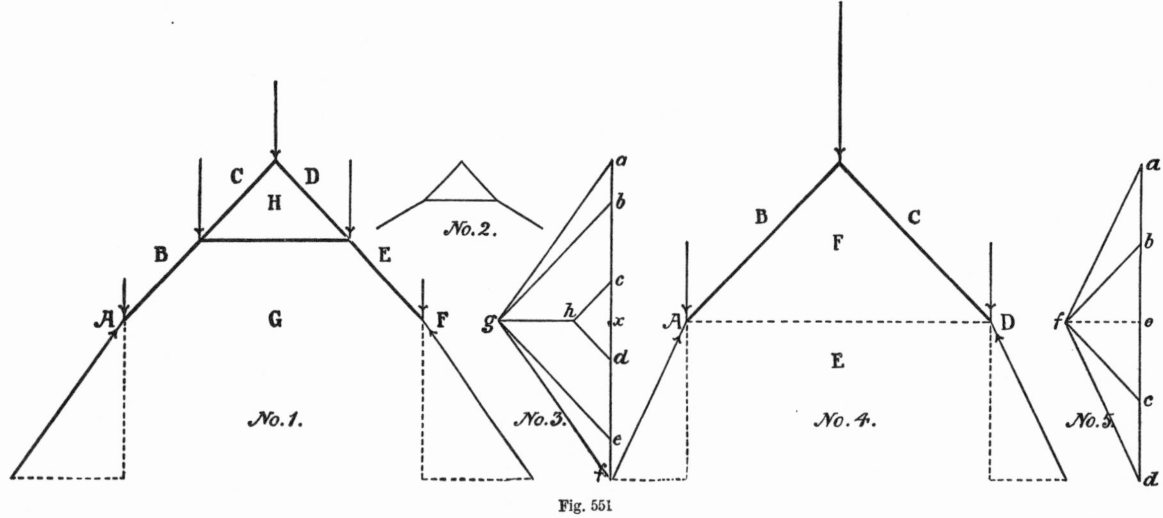

Fig. 551

at the apex. The five downward external forces are plotted at *a b c d e f*. For the joint at the apex we have the triangle of forces *c d h c*. For the joint at B C we can now obtain the polygon of forces *b c h g b*; the collar H G is shown to be in compression. Having obtained the stress *b g*, and the stress *a b* being known, the triangle of forces at A is completed by joining *a g*. The line *a g* gives us not only the magnitude, but also the direction of the supporting force at A, and this force has a vertical component $a x = \frac{W}{2}$, and a horizontal component or outward thrust $= x g$. It will be found that *x g* (fig. 551) = *g h* (fig. 550). In other words, the stress in the tie-beam is the measure of the outward thrust when the tie is omitted.

In practice, however, the rafters are rigid, and the collar H G is often intended to serve as a tie and prevent the spread of the truss. The ordinary graphic method of determining the stresses cannot therefore be applied to the case. The simplest method of procedure is to determine the outward thrust at the foot of the rafter, if the collar H G were omitted. This can be ascertained by drawing the stress diagram for the apex, as shown in Nos. 4 and 5, namely *b c f b*. The forces at the left-hand support will be *a b f a*, *f a* being the upward supporting force, having a vertical component $a e = \frac{W}{2}$ and a horizontal component *f e*. The outward thrust is therefore equal to *j e*, but this thrust is to be resisted by the tie placed at a higher level, and the stress in the collar will increase directly as its relative height is increased. In this case the collar is shown exactly midway between the feet and apex of the truss, and the stress in it will therefore be exactly twice as much as that obtained for the imaginary tie-beam F E. This is merely an application of the theory of moments; if x be the force required to resist the outward thrust at A, and y the vertical distance from A to the apex, then at the collar, as the vertical distance is $\frac{y}{2}$, the force required will be $2x$, for $x \times y = 2x \times \frac{y}{2}$.

Case 13.—*Butterfly or scissor-beam truss* (figs. 552–555). The form of truss shown in fig. 552 must be made with rigid rafters or it will collapse as shown in No. 2 (if the raking

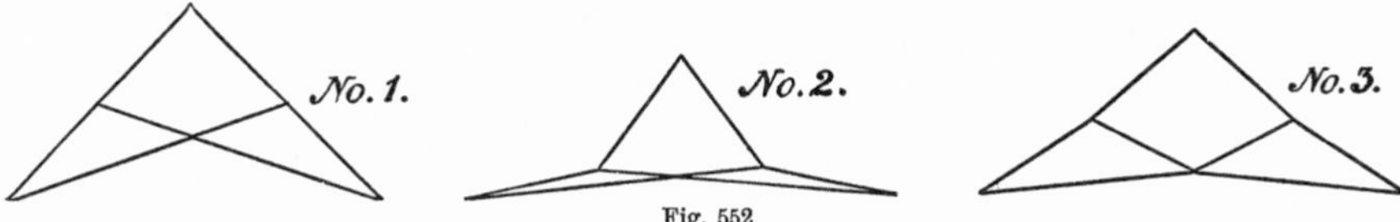

Fig. 552

ties are in one piece from end to end, and not fastened together at their intersection), or as shown in No. 3 (if the raking ties are jointed at the intersection). In the latter case the

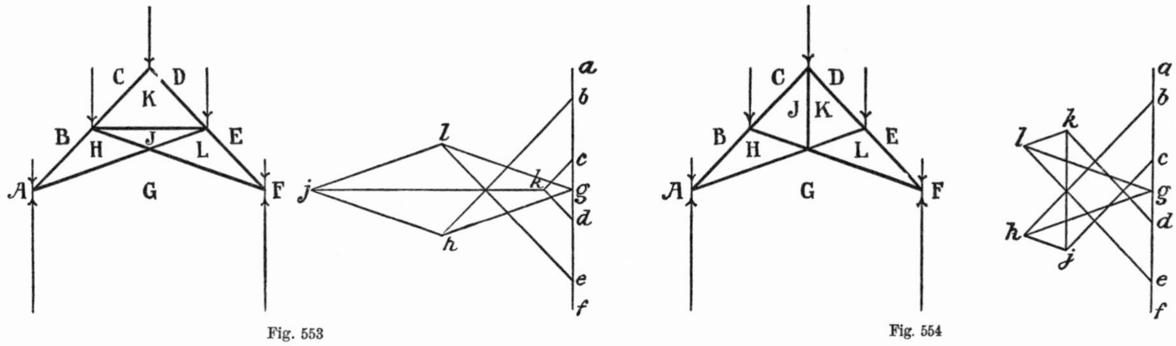

Fig. 553

Fig. 554

two triangles retain their original form, but the truss yields at their intersection and also at the three joints above them. It is impossible, therefore, to determine the stresses by the ordinary graphic method. To make the truss satisfactory, an additional member must be introduced in one of the three positions shown in figs. 553 to 555. The stress diagrams for these three varieties can all be drawn graphically, as shown in the illustrations, but it is not necessary to describe them in detail. Note that in fig. 553 K J is in compression, and J H, H G, G L, and L J in tension; in fig. 554 J K is in tension, as also are H G and G L, while J H and K L are in compression; and in fig. 555 L G is the only tension member.

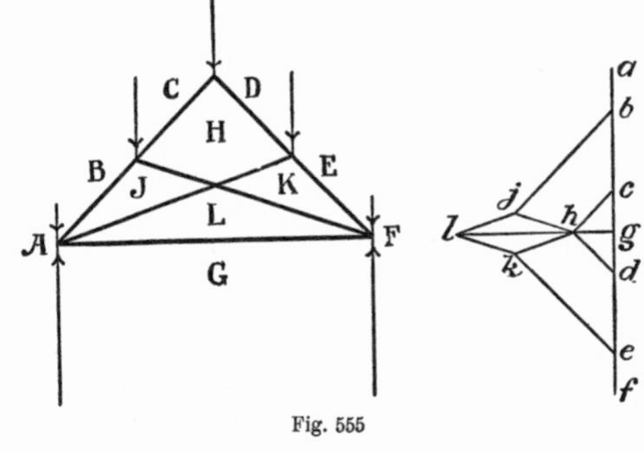

Fig. 555

Case 14.—*Hammer-beam truss* (fig. 556). The curved members perform exactly the same functions as if they were straight members extending from the feet of the truss to the foot of the king-post, as shown by the dotted lines, but of course the curve causes a cross-bending stress in addition to the direct stresses. This truss in practice almost invariably exerts a considerable outward thrust on the walls, on account of the slight yielding of the numerous joints. Theoretically, however, the truss is in equilibrium, and the stress diagram can be drawn as shown in fig. 556, No. 2. The wall-posts AM and KY, and the curved members ML and LY, are in this case assumed to be unnecessary,[1] and the spaces M and Y must therefore be considered merely as parts of L. The four forces at the joint ABNL will be $LA = \frac{W}{2}$, $AB = \frac{W}{16}$, and BN and NL, both unknown, and the stress polygon will be *labnl*. After ascertaining the stress diagram for the joint at BC, it will be necessary to proceed with the joint LNOP. The remaining parts can then be

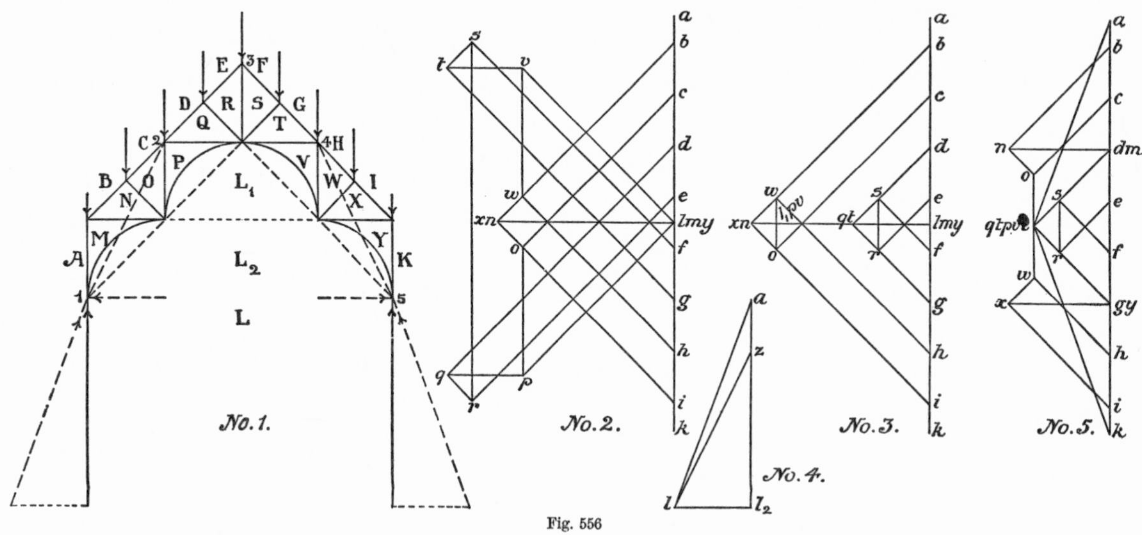

Fig. 556

drawn. It will be observed that the stresses in some of the members are exceedingly severe; the tensile stress in RS is practically equal to the total weight of the roof, and the compression in DQ and TG is almost as great.

The unscientific nature of the framing will be better appreciated by considering the stress diagram for the same roof with a tie-rod introduced from MN to XY, as shown by the dotted line. This tie-rod divides the space L into two parts, of which the upper part will be called L_1, and renders the curved members PL_1 and VL_1 unnecessary, except under an unsymmetrical load, such as wind. The truss becomes a combined king-and-queen-post truss, almost exactly like that in Case 9. The spaces P and V must be considered as parts of L_1. The stress diagram, No. 3, shows that the introduction of the tie-rod has reduced many of the stresses to a very large extent; thus, the stress in the king-post RS is only one-seventh of the stress shown for this member in No. 2, and the stresses in OP and VW (or OL_1 and VL_1) are only one-fifth of those in No. 2. The direction of some stresses is also altered by the introduction of the tie-rod—QP and TV in No. 2 are in tension, and in No. 3 in compression; OP and VW in No. 2 are in compression, and in No. 3 in tension; PL and VL in No. 2 are in tension, and in No. 3 are not stressed at all by a symmetrical load.

Thirdly, let the outward thrust of the truss at the points 1 and 5 be counteracted by buttresses. In this case the portion of the roof above the horizontal member 2–4 may be regarded as a separate framing supported on inclined framed struts, of which the lines

[1] In practice, however, they undoubtedly transmit part of the load by serving as brackets supporting the free ends of the hammer-beams, and in this way reduce the stresses in the upper curved members and in the king-post truss above. The stresses shown in No. 2 must therefore be regarded as the greatest stresses which can possibly be developed in the truss by the given load.

of resistance are 1–2 and 5–4 respectively. The vertical component of the supporting force at 1 is of course $\frac{W}{2}$, and the vertical downward force at the same point is half the load on 1–2 = $\frac{W}{8}$. The two stresses which have to be ascertained are therefore the outward thrust at 1 and the line of action of the supporting force. We will suppose the outward thrust to act as shown by the dotted line 1–5, the space above which will be called L_2, and the space between the vertical above 1 and the raking-strut 1–2 will be called Z. The polygon of forces at 1 (diagram No. 4) will therefore be *l a z l*, and the force $l l_2$ represents the outward thrust; *l a* is the line of action of the supporting force, and can be transferred to No. 1 as shown by the dotted lines. Assuming that the outward thrust is counteracted by suitable abutments, the truss will be in equilibrium, and the stress diagram can be drawn as shown in No. 5. It will be found that there are no stresses in the members Q P, P L, T V, and V L.

Under an unsymmetrical load, such as wind, different results will, however, be obtained in every case, and the curved members will be brought under stress.

WIND-PRESSURE.—In all calculations of the stresses in roof-trusses wind-pressure must be taken into consideration, as this often produces severer stresses than the dead loads. It is generally assumed that the pressure is normal to the inclination of the roof. The maximum pressure which must be allowed for is still a matter of dispute, but modern experiments show that it is not so great as was formerly supposed. In Mr. Baker's experiments, at the Forth Bridge, a pressure of 65 lbs. per square foot was recorded on one occasion, but it was found that the registering apparatus was faulty; this was corrected, and the greatest pressure recorded during the most violent gales of the next six years did not exceed 41 lbs. per square foot on the small gauge ($1\frac{1}{2}$ square foot in area), while the highest recorded pressure on the large gauge (20 feet by 15) was only 27 lbs. per square foot. The Board of Trade regulations for bridges in exposed situations require the calculations to be based on a wind-pressure of 56 lbs. per square foot, assumed to act *twice* over the exposed girder-surface.

It is obvious that in roofs the pressure normal to the surface will increase with the pitch of the roof. A high-pitched roof, therefore, not only exposes more surface to the wind, but receives a greater pressure per square foot. Some engineers assume that the wind-pressure in lbs. per square foot will be equal to the number of degrees of inclination of the roof-surface; thus, a pitch of 30° will be assumed to have a maximum wind-pressure of 30 lbs. per square foot. This is a simple rule, but over-estimates the pressure on roofs of greater pitch than 30° or 35°. Duchemin's formula, which is recommended by Professor Unwin, gives the following results, with a maximum horizontal wind-pressure of 40 lbs. per square foot:—

Inclination.	Wind-pressure in lbs. per square foot.	Inclination.	Wind-pressure in lbs. per square foot.
5°	6·89	45°	37·73
10°	13·59	50°	38·64
15°	19·32	55°	39·21
20°	24·24	60°	39·74
25°	28·77	65°	39·82
30°	32·00	70°	39·91
35°	34·52	75°	39·96
40°	36·40	80° to 90°	40·00

Two peculiarities must be noticed in regard to wind-pressure: (1) its action is assumed to be at right angles to the slope of the roof; (2) it acts on one side only of the ordinary gabled roof, and consequently produces unsymmetrical loading of the trusses.

To find the stresses produced by wind-pressure on roof-trusses, the same principles are involved as in the dead-load diagrams already given, and it will be sufficient therefore for us to consider two or three typical cases. In carpentry the ends of roof-trusses are generally fixed, and not free to move on rollers as in many steel and iron roofs, and only trusses with fixed ends need therefore be considered here. In these roofs the supporting forces are generally assumed to be normal to the windward rafter.

Case 15.—*Wind-pressure on king-post truss.* An example of this is given in Plate XXV. Figs. 1 and 2 show the ordinary dead-load diagrams for a roof without ceiling; figs. 3 and 4 show the loads and stresses given by a ceiling weighing 10 lbs. per square foot and a roof-covering weighing 16 lbs. per square foot. These diagrams do not introduce any new features, but are drawn to a larger scale than has been adopted in the illustrations. In fig. 5 the wind-pressure on the left-hand side of the roof is taken as being concentrated at the three joints, the magnitude, direction, and point of application being shown by the full lines at right angles to the rafter. The resultant of these three parallel forces is the line 1–2, and the reactions at the supports can be found graphically by drawing lines from A and F parallel to 1–2; from the extremity 2 draw the line 2–3 parallel to A F, and cutting A 3 in 3; join 3 F by the line cutting 1–2 in 4. The line 1–2 is divided into two parts, namely, x_1 representing the reaction at A, and y_1 representing that at F. The reactions at the supports having been found, the stress diagram (fig. 6) can be drawn without difficulty. If the joints do not divide the truss into equal parts, the funicular polygon must be drawn in order to obtain the reactions, as explained on page 303. Fig. 7 shows the stress diagram obtained by finding the vertical components of the three forces, as shown by the dotted lines above the left-hand rafter in fig. 5; the horizontal components are not considered, and the stress diagram is not as correct as fig. 6.

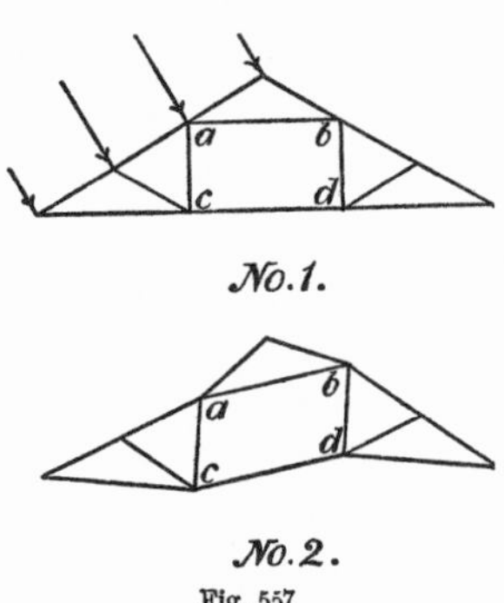

Fig. 557

Case 16.—*Wind-pressure on queen-post truss* (figs. 557 and 558). The stress diagram for the dead load on a truss of this form was given in fig. 547, and it follows that under a symmetrical load the truss is in equilibrium. But when we attempt to draw the stress diagram for an unsymmetrical load, like that due to wind, we find that it cannot be done, and the conclusion is forced upon us that cross-bending stresses must be set up in some of the members. In

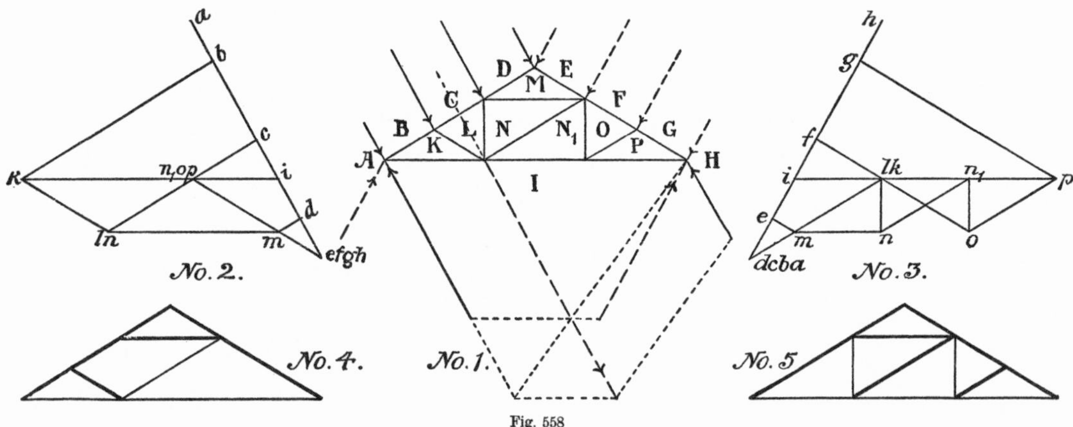

Fig. 558

fig. 557, No. 1 represents such a truss with a wind-pressure on the left-hand rafter; it is obvious that if the four joints, *a*, *b*, *c*, and *d* are hinged, the truss will yield under the unsymmetrical load into a shape like that shown in No. 2. The queen-post truss, therefore, while stable under the ordinary dead load, is not well adapted for resisting wind-pressure, as this causes a bending stress in the rafters or tie-beam. If a member is introduced diagonally

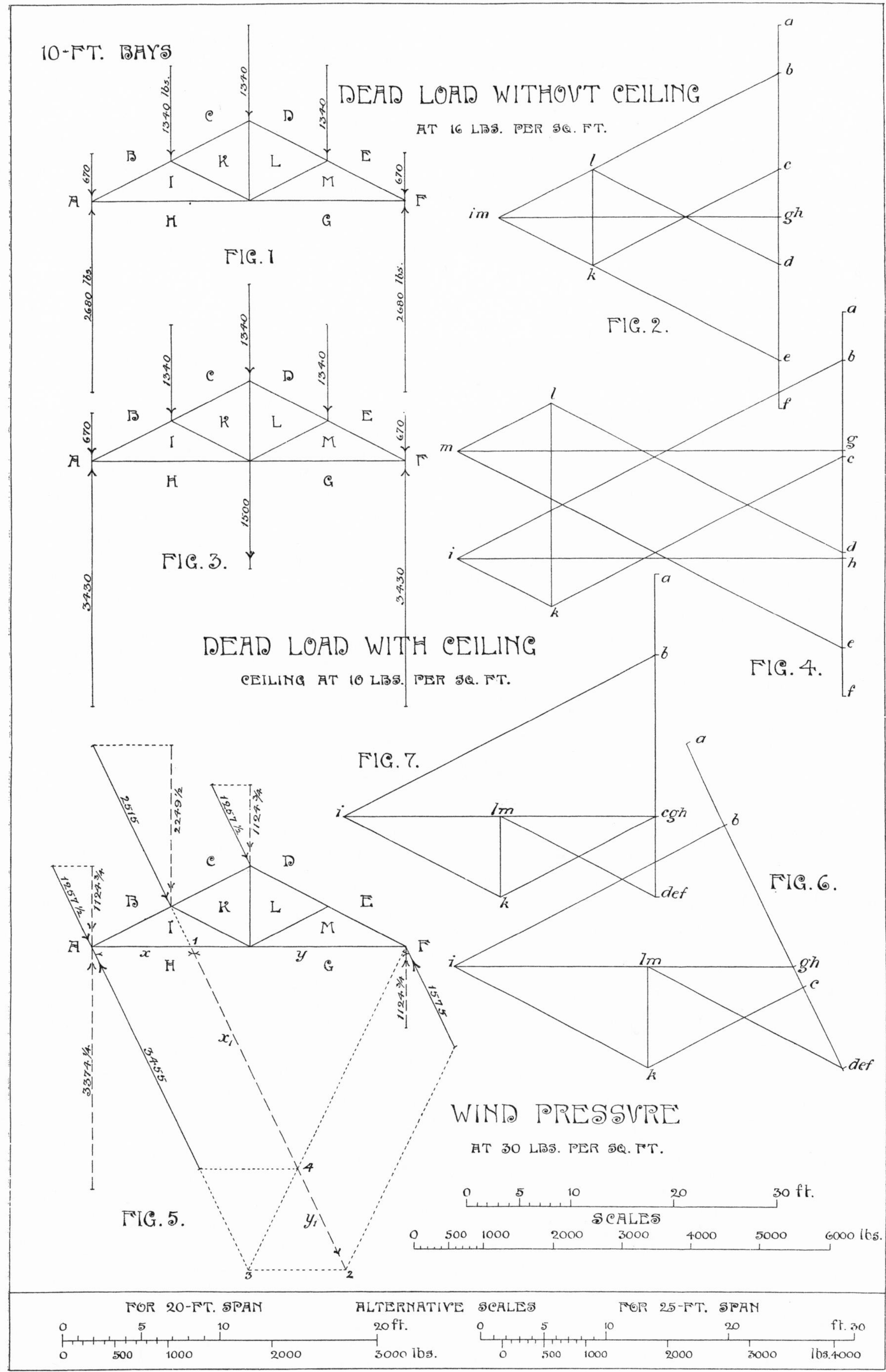

STRESS DIAGRAMS FOR KINGPOST TRUSS

across the central oblong in the truss, as shown in fig. 558, the truss will be properly braced, and the stress diagrams can be drawn in the usual manner. The stress diagram for a wind on the left-hand side of the roof is shown in No. 2, and that for a wind on the right-hand side in No. 3. In both cases both rafters are in compression, and the whole of the tie-beam is in tension. The stresses in the other members are as follows—compression being represented by a + sign, and tension by a − sign; where no sign occurs the member is without stress:—

Wind on the left	+ K L, + M N,	L N,	− N N_1,	N_1 O,	O P.
Wind on the right	K L, + M N,	− L N,	+ N N_1,	− N_1 O,	+ O P.

The stressed members are shown in Nos. 4 and 5, those in compression by thick lines, and those in tension by fine lines.

When the wind is on the left there are no stresses (except, of course, those due to the dead load) in L N, N_1 O, and O P, and N N_1 is in tension: when the wind is on the right there is no stress in L K, and N N_1 is in compression. N N_1 will therefore be a strut or a tie according to the direction of the wind, and must be designed accordingly. What alteration in the stresses would result from the introduction of another diagonal member from D C L M to I N_1 O P may be left to the reader's ingenuity.

Single-stress diagrams for dead load and wind-pressure.—The variation in the stresses produced in different members of a truss by dead load and wind-pressure points to the conclusion that the most satisfactory results will be obtained by considering the two together. One example will be given to show the method of procedure.

Case 17.—*Combined stress diagram for king-post truss* (fig. 559). We will assume that the wind-pressure normal to the roof is exactly double the vertical pressure due to the dead

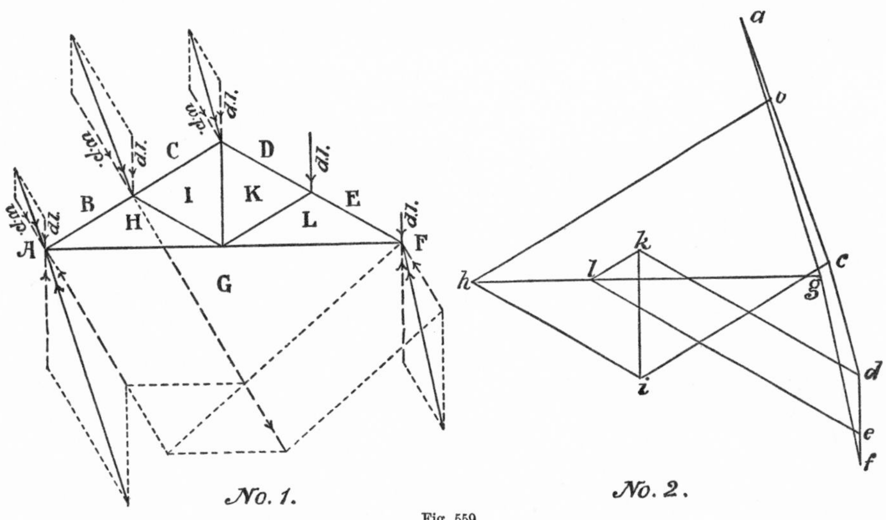

Fig. 559

load at each of the joints on the left-hand rafter. On the other rafter there will be only the dead load. The forces at each joint must be plotted as shown, *w.p.* = wind-pressure, and *d.l.* = dead load. The resultant of these forces gives the combined pressure in magnitude and direction. The reactions at the supports must be obtained in a similar manner. The stress diagram can then be drawn without difficulty.

Case 18.—*Warren girder* (fig. 560). The Warren girder may be used as a roof or bridge truss. It consists of top and bottom chords with oblique struts and ties. In the example illustrated the truss is supposed to be loaded uniformly along the top chord and outermost struts by the roof covering, and uniformly along the bottom chord by the ceiling. The downward force A B is composed of half the loads on B O and O N; and G H of half those on G X and X I. The stress diagram presents no difficulty. It will be found that the compression members are those shown by thick lines, the other members being in tension.

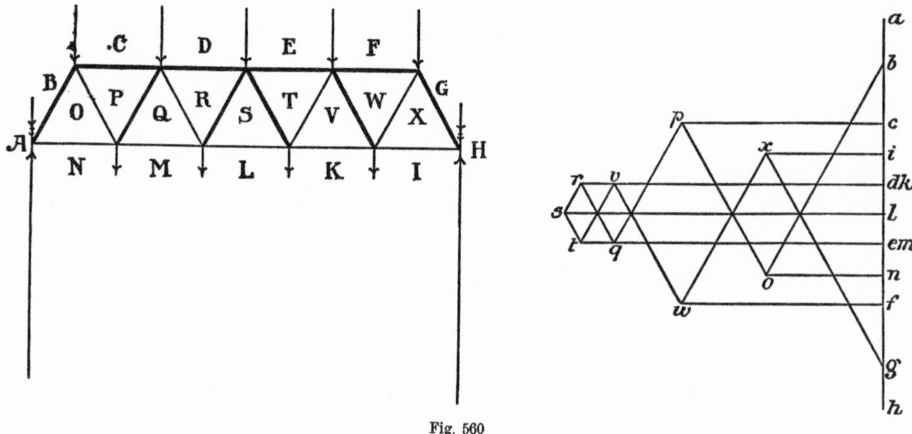

Fig. 560

Case 19.—*Howe truss* (fig. 561). In this the ties are vertical and the struts oblique. The Pratt truss is of similar design, but with vertical struts and oblique ties. The loads due to the roof covering and ceiling are (in the example illustrated) supposed to be uniformly distributed, but as the members are not of equal length, the weights concentrated at the

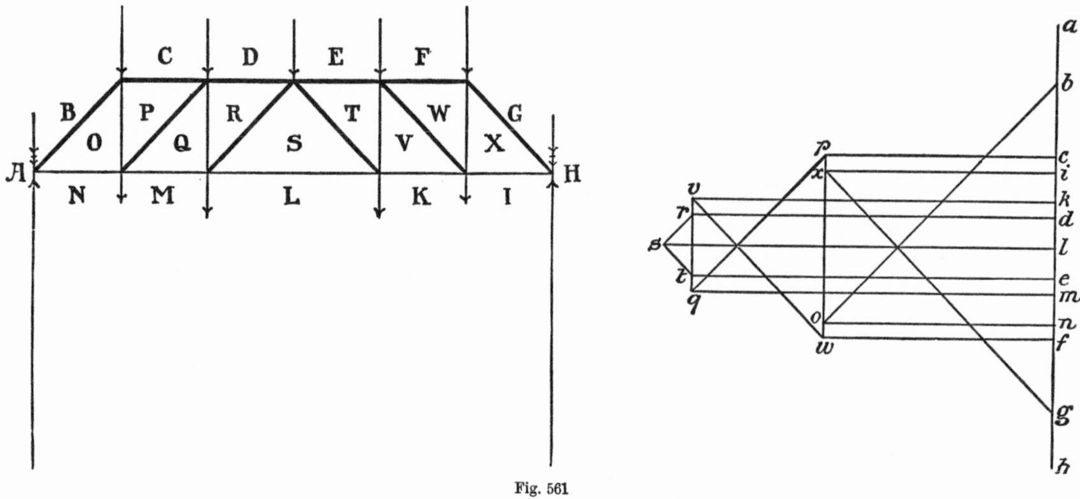

Fig. 561

joints are not equal; thus B O is longer than D R, and B C is therefore greater than C D; and S L is longer than O N, and L M is therefore greater than M N. On drawing the stress diagram it will be found that the compression members of the truss are those shown by thick lines, the others being in tension.

In both the Howe and Warren trusses the stresses in the top and bottom chords (under uniform loads) increase from the end bays to the centre, while the stresses in the struts and ties increase from the centre of the truss to the ends.

In a general work of this kind it is impossible to consider a tithe of the cases which occur in practice, but sufficient has perhaps been said to give the student some idea of the graphic method of determining the stresses in framed structures, and to enable him to discover the parts which the several members may be called upon to play. For further information, books dealing specially with the subject must be consulted.

PART II.—PRACTICAL

CHAPTER I

PHYSICAL AND MECHANICAL PROPERTIES AFFECTING THE STRENGTH OF TIMBER

No one who has examined a series of tests of timber, or who has compared the tests carried out on any variety of timber by different persons, can have failed to be struck with the wide divergence in the strength of many of the specimens. It is this divergence which has rendered all attempts to ascertain the normal strength of timber more or less unsuccessful. The important investigations instituted by the U.S.A. Division of Forestry have added much to our knowledge of the subject, but unfortunately they were abandoned while still incomplete. Nevertheless, the published records of these investigations furnish the most trustworthy information which has yet been obtained regarding North American timbers, and they will therefore be largely quoted in this chapter.

Among the most important properties affecting the strength of timber are—(1) The percentage of moisture or degree of seasoning, and (2) the specific weight of the dry timber. Other factors which require consideration are—(3) The method of seasoning, (4) the part of the tree from which the timber is cut, (5) the age of the tree, (6) the nature of the "annual rings", and (7) the presence of knots, shakes, &c.

1. THE PERCENTAGE OF MOISTURE IN TIMBER

It has long been known that "green" timber is weaker than seasoned. Sixty years ago Hodgkinson made a few tests for the purpose of ascertaining the difference of strength, but his experiments were not carried out with sufficient accuracy to render the results of much use. Bauschinger's tests are more satisfactory, and show that "green" pine timber may have less than half the compressive strength of dry, but that the difference is not so great for transverse stresses. The results of his compressive tests on "Scotch pine" are plotted in fig. 562. Four pieces of timber were taken, and sections from these were cut and tested "green"; these, marked A in the figure, contained on the average 59 per cent of moisture (calculated on the *dry* weight of the wood) or 37 per cent (calculated on the *wet* weight). Four tests (marked B) were afterwards made on partly-seasoned sections; these contained on the average moisture equal to 17 per cent of the "dry" weight or 14·6 per cent of the "wet" weight. Finally, four tests (marked C) were made on well-seasoned sections, containing 9 per cent of moisture (on the "dry" weight) or 8·2 per cent (on the "wet" weight). The partly-seasoned wood was nearly half as strong again, and the wood thoroughly seasoned about two-and-one-third times as strong as the green wood.

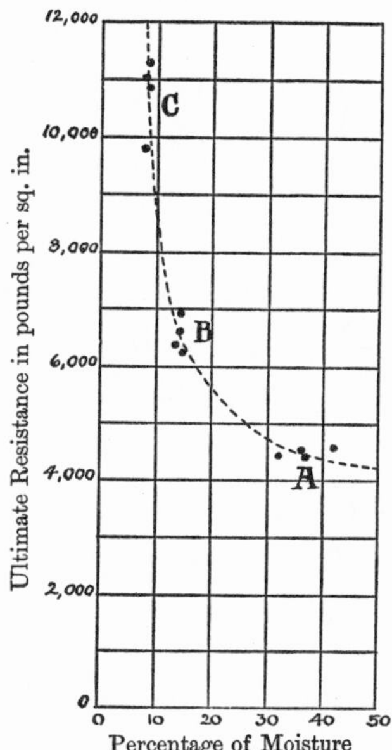

Fig. 562.—Compressive Strength of Scotch Pine. A, "green"; B, partly-seasoned; C, well-seasoned. (After Bauschinger.)

The experiments carried out by Professor J. B. Johnson and Mr. Filibert Roth for the U.S.A. Division of Forestry are particularly valuable, on account of the number of tests which were made and the care with which the investigations were conducted. The relative values of the tests on the four southern pines,[1] commonly known as "yellow" pine in America and as "pitch" pine in this country, are given in Table I. The results are stated to be the "average of all valid tests".

Table I.—INFLUENCE OF MOISTURE ON STRENGTH OF SOUTHERN PINES

Nature of Stress.	Percentage of moisture calculated on "dry" weight.	Relative Strength.[2]				
		Cuban.	Long-leaf.	Loblolly.	Short-leaf.	Average.
Bending	33 per cent[3]	100	100	100	100	100
	20 "	118	116	117	118	117
	15 "	142	142	138	134	139
	10 "	181	182	168	160	173
Crushing endwise ...	33 "	100	100	100	100	100
	20 "	132	122	128	122	126
	15 "	157	154	156	142	152
	10 "	184	206	206	168	191
Mean of both bending and crushing	33 "	100	100	100	100	100
	20 "	125	119	122	120	122
	15 "	149	148	147	138	146
	10 "	182	194	187	164	182

These tests give practically identical results with those of Bauschinger, within the same range of moisture. On the average the strength of the timber increases 46 per cent by "ordinary good yard seasoning", and a further 36 per cent by "complete seasoning in kiln or house". Mr. B. E. Fernow, chief of the Division of Forestry, states, in regard to these tests: "Large timbers require several years before even the yard-season condition is attained, but 2-inch and lighter material is generally not used with more than 15 per cent moisture".

The importance of these tests can scarcely be over-estimated. They show clearly that little reliance can be placed on tests of timber unless the amount of moisture has been carefully ascertained. The failure to determine this factor is one cause of the discrepancy between the breaking-weights recorded by different experimenters. It also accounts in a great measure for the difference in strength observed between small test-pieces and large scantlings of the same timber.

The seasoning of timber has a twofold influence: it causes the wood as a whole to shrink, and so brings about 10 per cent more fibres into a given area of cross-section; and it also contracts, and thereby hardens the cell-walls.

2. THE SPECIFIC WEIGHT OF DRY TIMBER

The specific gravity or weight of dry timber is another important factor of strength. It has long been known that, of timber of the same species, the heavier is generally the stronger, and recent experiments appear to indicate that this truth is of wider application. Bauschinger's tests of pine, larch, spruce, and fir led him to the conclusion that the crushing-strength of these timbers increases almost directly as the specific gravity, or, more exactly, crushing-strength (in lbs. per square inch) = (13800 × specific gravity) − 900.

[1] For a description of these woods, see pp. 91–92.

[2] The strength of timber with 33 per cent of moisture being taken as 100 in each case, as higher degrees of moisture do not appear to alter the strength.

[3] 33 per cent is "green" timber, 20 per cent "half-dry", 15 per cent "yard-dry", and 10 per cent "room-dry".

American tests of "green" cypress, whose specific weight varied from ·58 to ·35, showed a gradual diminution of strength from 6500 to 2900 lbs. per square inch under compression, and from 7800 to 5000 under a transverse stress.

The specific gravity or weight of the four southern pines already described is a very accurate index of their transverse and compressive strength, but apparently the relationship between the weight and the tensile and shearing strength is not so intimate, at any rate as regards Cuban pine. The relative values are given in Table II, the value of Cuban pine being taken at 100, except for tension and shearing, the values for which are worked out from the weight-value of loblolly pine (84·1).

TABLE II.—SPECIFIC GRAVITY AND STRENGTH[1] OF SOUTHERN PINES

	Cuban.	Long-leaf.	Loblolly.	Short-leaf.
Specific gravity of "dry"[2] wood ...	0·63	0·61	0·53	0·51
Weight per cub. ft. in lbs.	39	38	33	32
Relative specific gravity	100	96·8	84·1	80·9
Relative transverse strength ...	100	91·2	84·5	77·2
„ compressive „	100	87·2	82·8	75·1
„ tensile „	83·5	88·7	84·1	78·2
„ shearing „	82·8	86·0	84·1	83·8

In connection with this table it must be pointed out that the relative strengths are the averages of all the tests, while the values of the specific gravity have been computed from a comparatively small number of specimens. A series of specimens tested both for transverse strength and specific gravity showed a remarkable relationship; taking the value of *Cuban pine* at 100 both for weight and strength, we have, for *Long-leaf*, strength = 91, s.g. = 94; for *Loblolly*, strength = 84, s.g. = 82; and for *Short-leaf*, strength = 77, and s.g. = 77.

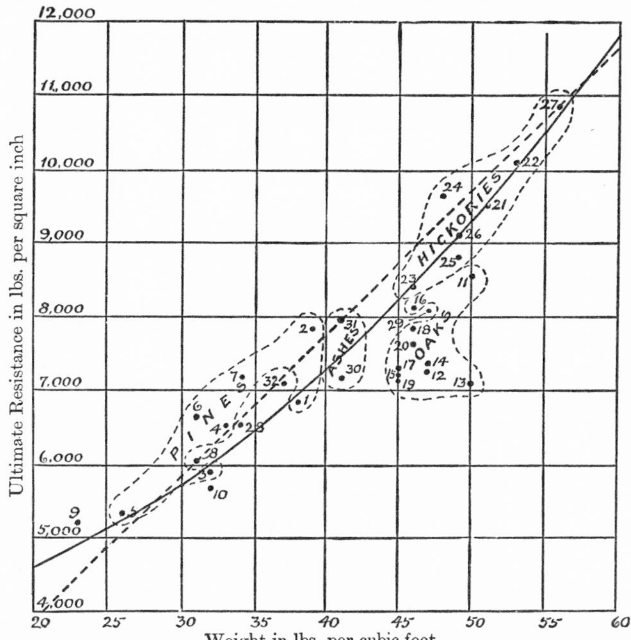

Fig. 563.—Relation of Compressive Strength to Weight of Timber. (The number at each point refers to the species of timber, a list of which is given in Table III.)

A more important point for determination is to ascertain whether the relationship between weight and strength holds good for timbers of different species and genera. The conclusions of the U.S.A. Division of Forestry are cautiously stated as follows:—"We can now extend the application of the law of relation between weight and strength a step farther, and state as an indication of our tests that probably, *in woods of uniform structure, strength increases with specific weight,* independently of species and genus distinction, i.e. *other things being equal, the heavier wood is the stronger.* We are at present inclined to state this important result with caution, only as a probability or indication."[3]

[1] "The values of strength refer to all tests, and therefore involve trees of wide range of age and consequently of quality,—especially those of long-leaf involve much wood of *old* trees,—hence the relation of weight and strength appears less distinct."—*Circular No. 12, U. S. Dep. of Agriculture, Division of Forestry.*

[2] That is to say, wood with less than 15 per cent moisture.

[3] *Circular No. 15, U. S. Dep. Agriculture, Div. Forestry.*

The results of the tests on thirty-two kinds of timber are shown graphically in figs. 563 and 564; the numbers adjoining the tests refer to the species of timber, a list of which, together with the specific gravity and the number of tests applied to each species, is given in Table III. Where the number of tests is small, the values for strength and weight must be regarded as approximate only.

The report goes on to state that an exception is apparent in the oaks, and that this is probably due to "the structure of oak wood being exceedingly complicated and essentially different from that of the wood of all other species under consideration". It is clear that this exception is not warranted by the tests; if a straight line is drawn between oaks No. 17 and 19 and through hickory No. 21 (fig. 564), it divides the tests of these woods into two equal parts, three hickories and five oaks lying on each side of it. In other words, hickory and oak appear to behave in a very similar manner, a slight increase of weight in each being (on the average) accompanied by a considerable increase of strength.

The relationship between strength and weight may (perhaps more correctly) be expressed graphically by a curved line, and not by the straight dotted line marked in the figs. Such a curved line is shown at AA, that for transverse strength being calculated from an empirical formula, devised by the writer—

$$W = 5000 + 3{\cdot}8\, w^2 \qquad (13)$$

Where W = transverse strength in lbs. per sq. in., and w = weight of a cub. ft. of "dry" wood in lbs.

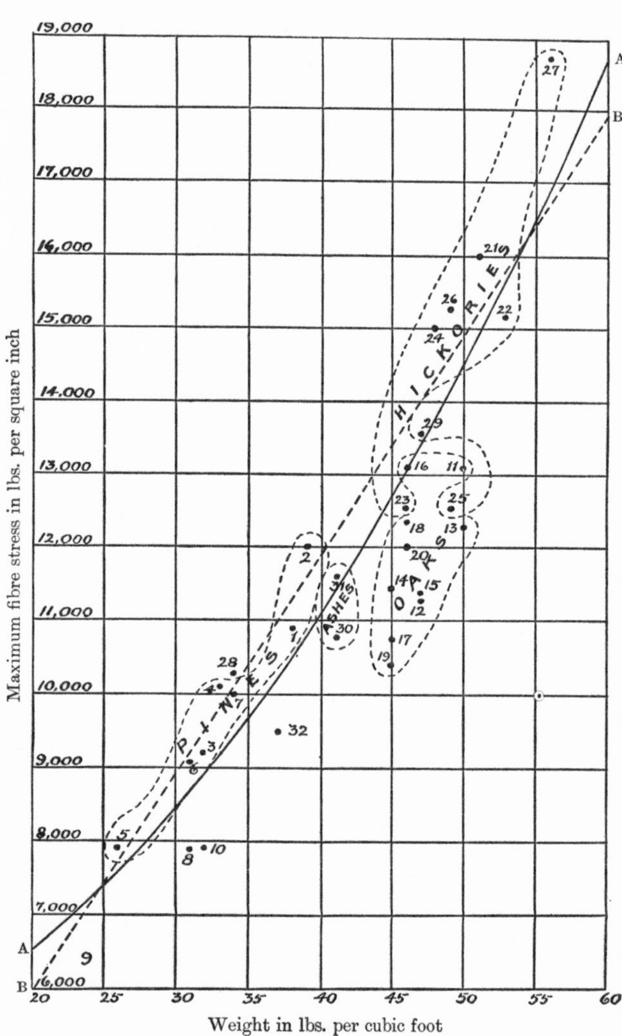

Fig. 564.—Relation of Transverse Strength to Weight of Timber. (The number at each point refers to the species of timber, a list of which is given in Table III.)

This curve gives a better average of all the tests than the straight line BB. Sixteen woods lie on each side of the curve, while the straight line has eight woods on one side and twenty-four on the other.

It is unnecessary, however, to pursue the enquiry further, as more elaborate tests may give different results. The main point is that specific gravity has a most important influence on strength, although this influence will be more or less affected by other conditions, such as moisture, method of seasoning, knots and cracks, and the general structure of the wood. Tests of small scantlings of timbers from Jamaica and British Guiana, carried out by Professor Unwin, show that these conditions affect the strength to a serious extent. In the main, the heavier woods were stronger in every respect than the lighter woods, but some timbers were exceptions to the rule, being either light and strong or heavy and weak.[1]

To ascertain the specific gravity of wood, special apparatus is required, but a close approximation is possible (in certain kinds of timber) by simple inspection or measurement. The greater the relative proportion of the dark summerwood rings, the greater will be, other things being equal, the specific weight

[1] See *Imperial Institute Journal*, July, 1896, and April, 1897.

and the strength. Thus, in the southern pines, the dark rings have, in good timber, a specific weight of from ·9 to 1, while the light-coloured springwood rings have a specific weight of about ·4 only; a log, therefore, in which the dark rings are one-third of the whole, will have a specific weight of $(\frac{1}{3} \times \cdot 9) + (\frac{2}{3} \times \cdot 4) = \cdot 56$. In the words of Mr. B. E. Fernow, "The relative amount of summerwood furnishes altogether the most delicate and accurate measure of these differences of weight as well as strength, . . . especially since this relation is free from the disturbing influence of both resin and moisture contents of the wood, so conspicuous in weight determinations".

TABLE III.—SPECIFIC GRAVITY, &c., OF THIRTY-TWO AMERICAN WOODS TESTED BY THE U.S. DIVISION OF FORESTRY

No.	Name of Species.	Number of Trees.	Number of Mechanical Tests.	Average Specific Gravity of Dry Wood.	Localities where Specimens were grown.
1	Long-leaf Pine (*Pinus palustris*)	68	6478	·61	Alabama, Georgia, South Carolina, Mississippi, Louisiana, and Texas.
2	Cuban Pine (*Pinus cubensis* or *heterophylla*)	12	2113	·63	Alabama, Georgia, and South Carolina.
3	Short-leaf Pine (*Pinus echinata*)	22	1831	·51	Alabama, Missouri, Arkansas, and Texas.
4	Loblolly Pine (*Pinus tæda*)	32	3335	·53	Alabama, Arkansas, Georgia, and South Carolina.
5	White Pine (*Pinus strobus*)	17	540	·38	Wisconsin and Michigan.
6	Red Pine (*Pinus resinosa*)	8	412	·50	Wisconsin and Michigan.
7	Spruce Pine (*Pinus glabra*)	4	696	·44	Alabama.
8	Bald Cypress (*Taxodium distichum*) ...	20	3396	·46	South Carolina, Louisiana, and Mississippi.
9	White Cedar (*Chamæcyparis thyoides*) ...	4	354	·37	Mississippi.
10	Douglas Spruce (*Pseudotsuga taxifolia*) ...	...	225	·51	Lumber yard.
11	White Oak (*Quercus alba*)	12	1009	·80	Alabama and Mississippi.
12	Overcup Oak (*Quercus lyrata*)	10	911	·74	Mississippi and Arkansas.
13	Post Oak (*Quercus minor*)	8	256	·80	Alabama and Arkansas.
14	Cow Oak (*Quercus michauxii*)	11	935	·74	Alabama, Arkansas, and Mississippi.
15	Red Oak (*Quercus rubra*)...	7	299	·73	Alabama and Arkansas.
16	Texan Oak (*Quercus texana*)	3	479	·73	Arkansas.
17	Yellow Oak (*Quercus velutina*)	5	222	·72	Alabama.
18	Water Oak (*Quercus nigra*)	4	132	·73	Mississippi.
19	Willow Oak (*Quercus phellos*)	12	649	·72	Alabama, Arkansas, and Mississippi.
20	Spanish Oak (*Quercus digitata*)	11	1035	·73	Alabama, Arkansas, and Mississippi.
21	Shagbark Hickory (*Hicoria ovata*) ...	6	794	·81	Mississippi.
22	Mockernut Hickory (*Hicoria alba*) ...	4	300	·85	Mississippi.
23	Water Hickory (*Hicoria aquatica*) ...	2	197	·73	Mississippi.
24	Bitternut Hickory (*Hicoria minima*) ...	4	100	·77	Mississippi.
25	Nutmeg Hickory (*Hicoria myristicæformis*)	3	294	·78	Mississippi.
26	Pecan Hickory (*Hicoria pecan*)	2	172	·78	Mississippi.
27	Pignut Hickory (*Hicoria glabra*) ...	3	84	·89	Mississippi.
28	White Elm (*Ulmus americana*)	2	91	·54	Mississippi.
29	Cedar Elm (*Ulmus crassifolia*)	3	201	·74	Arkansas.
30	White Ash (*Fraxinus americana*) ...	3	476	·62	Mississippi.
31	Green Ash (*Fraxinus lanceolata*) ...	1	45	·62	Mississippi.
32	Sweet Gum (*Liquidambar styraciflua*) ...	7	508	·59	Arkansas and Mississippi.

3. THE METHOD OF SEASONING

We have seen that seasoned timber is much stronger than green, but the method of seasoning requires consideration, as it has a considerable effect on the ultimate strength of timber. Many experts are of opinion that natural seasoning gives the best results, but the time required is so long that more rapid methods are now preferred. Drying in kilns is the process most generally adopted, and, if due care is exercised, appears to be perfectly satisfactory; on the other hand, if carried on too quickly, it may prove injurious.

"Rapid drying of the heavier hardwoods of complicated structure, especially in large sizes and from the green state, is apt to produce inordinate checking [*i.e.* cracking], and thus weakening of the material."[1] For other woods, however, kiln-drying may be quite as

[1] *Circular No. 12, U. S. Dep. Agriculture, Div. Forestry.*

satisfactory as natural seasoning. Cuban pine in small scantlings (4 inches by 4 inches) has been found to have the same transverse and compressive strength when seasoned in warm air (about 100° F.) and when dried at temperatures varying from 150° at the entrance of the kiln to 190° at the exit.

Higher temperatures, however, appear to be injurious. Scantlings of long-leaf pine were tested after drying at a temperature of over 300° F., and under a pressure of 150 lbs. The process was found to reduce the compressive strength of the material about 24 per cent, and the transverse strength no less than 37 per cent.

The soaking of wood by rafting or floating does not appear to have any injurious effect on the strength, provided that the wood is properly dried or seasoned before use.

4. OTHER CONDITIONS AFFECTING THE STRENGTH OF TIMBER

Different parts of the same tree vary greatly in strength, as well as in specific weight. This is shown graphically in fig. 565, which is a diagrammatic vertical section of the trunk of a long-leaf pine reproduced from the Circular No. 12 already referred to. It will be observed that the lightest and therefore weakest part of the tree is the pith and the adjacent rings, representing the young sapling of less than fifteen or twenty years' growth. The maximum weight is grown between the ages of forty and sixty years. In other varieties of timber the age of greatest weight and strength is different; thus, in loblolly and short-leaf pines the maximum is reached between the ages of thirty and forty years.

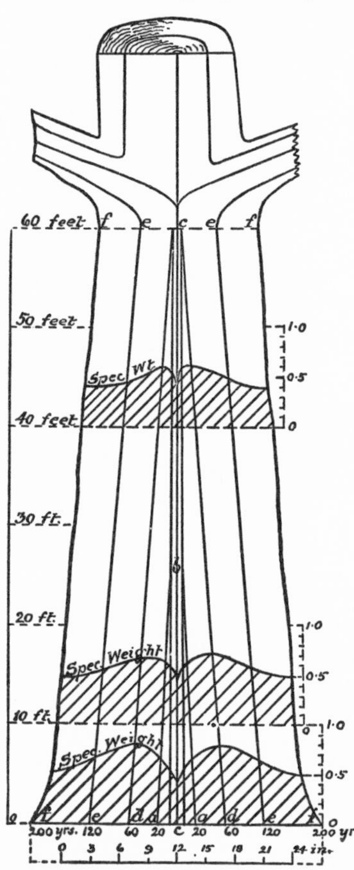

Fig. 565.—Diagrammatic Vertical Section through Stem of Long-leaf Pine, showing Variation of Specific Weight with Height, Diameter, and Age. *aba* =tree 20 years old, *dcd* = 60 years, *eeee* =120 years, and *fff* = 200 years.

The diagram shows clearly the weakness of the sapwood of old trees, but this weakness is not due to the sap but to the fact that in old age trees naturally produce lighter and weaker wood. The sapwood of a tree about forty or sixty years old may be as strong as the heartwood, or even stronger.

The weight and strength decrease from the butt or stump upwards. This is shown by the shaded portions in fig. 565. It is proved by a series of experiments, part of which are summarized in Table IV. The weight and strength of long-leaf timber cut from the tree at the height of 60 feet from the ground are about 24 or 25 per cent less than of wood cut from the stump.

Of course, these statements about the difference of strength in different parts of a tree refer only to the clean portions of the wood, and are "variably affected at each whorl of knots (every 10 to 30 inches) according to their size, and also by the variable amounts of resin (up to 20 per cent of the dry weight)".

The nature of the soil and climate has some influence on the strength of timber, but it is much more marked in some varieties than in others. Writing in 1892, Mr. B. E. Fernow stated that "such 'ring-porous' woods as the oaks and ash show the greatest strength and elasticity when their annual rings are wide [*i.e.* when the wood is quickly grown], while the slow-grown mountain oak seems to excel in stiffness. From conifers, on the other hand, according to Hartig, the slow-grown timbers seem to exhibit superior qualities; hence those from rich soils are not desirable. This again has appeared doubtful, or at least true only within unknown limits, from Bauschinger's experiments, who showed that tensile strength in

TABLE IV.—SPECIFIC WEIGHT AND STRENGTH OF LONG-LEAF PINE AT DIFFERENT HEIGHTS IN THE TREE

No. of Feet from Stump.	Specific Weight.		Transverse Strength.		Compressive Strength.	
	Absolute.	Relative.	Absolute in lbs. per sq. in.	Relative.	Absolute in lbs. per sq. in.	Relative..
0	·751	*106*	......	...	......	...
6	·705	*100*	12,100	*100*	7350	*100*
10	·674	*96*	11,650	*96*	7200	*98*
20	·624	*89*	10,700	*88*	6800	*93*
30	·590	*84*	10,100	*84*	6500	*89*
40	·560	*80*	9,500	*79*	6300	*86*
50	·539	*77*	9,000	*75*	6150	*83*
60	·528	*75*	8,600	*71*	6050	*82*

"Logs from the top can usually be recognized by the larger percentage of sapwood and the smaller proportion and more regular outlines of the bands of summerwood, which are more or less wavy in the butt-logs."

pines was independent of the total width of the annual ring, but dependent on the ratio between the spring wood and summer wood. That wet soil produces brittle, [and] dry or fresh soil tough timber, is believed, but needs proof." Later investigations by Mr. Fernow and his co-workers show that a sweeping generalization on the subject is impossible. Thus, locality—which includes both soil and climate—"appears to have little influence" on the weight and strength of Cuban and Long-leaf pines, but has a material influence on Short-leaf pine,—"the wood from the southern coast and Gulf region, and even Arkansas, is generally heavier than the wood from localities farther north; very light and fine-grained wood is seldom met near the southern limit of the range [of growth], while it is almost the rule in Missouri, where forms resembling the Norway pine are by no means rare". Loblolly pine is also influenced by soil and climate, but to a smaller extent. In regard to cypress, the investigations in 1898 showed that "difference in quality of material is evidently far more a matter of individual variation than of soil or climate"; and again, "the emphatic claim recently made by European writers that good soil and heavy timber universally go together, has never been so perfectly contradicted as by these experiments, which show that some of the Pond cypress from South Carolina, grown on the very poorest soil and requiring four centuries to barely make a telegraph pole, is both the heaviest and strongest material of the collection".

With reference to Australian timbers, Mr. J. Ednie-Brown, Conservator of Forests for Western Australia, writes: "It is well known here that soil, locality, temperature, elevation, and rainfall affect the strength and durable qualities of timbers. In no case is this more forcibly demonstrated than in the practical use of our best commercial timbers, the karri and jarrah. For instance, it is well known that jarrah . . . grown upon the ironstone formations or ridges, is far superior in every respect to that grown upon the low-lying granitic soils, although the trees themselves, and the appearance of the timber as well, may present the same characteristics to the eye. So with karri; I have seen specimens which have been in the ground over thirty years in good condition, whilst others grown in a different locality have only lasted a few years under similar conditions."[1]

Bauschinger found a marked difference in the strength of whitewood or spruce from different localities, as shown in the following table, which is also interesting for the light it sheds on the influence which the date of felling has on the strength of timber.

The tests three months after felling show that the winter-felled wood is much stronger than summer-felled, but this advantage is lost with thorough seasoning, as it is merely due to the fact that in winter there is less moisture in the wood than in summer. The tests of the seasoned wood (columns 2) may be compared by averaging them, when it will be found

[1] Second and Amended Report upon *The Forests of Western Australia and their Development*, 1899.

that the average strength of the summer-felled timber is 6114 lbs., and of the winter-felled 6055, so that the time of felling has no permanent influence on the strength of the timber.

Knots, cracks, and other defects affect the strength of timber very injuriously, nearly all columns between about 5 and 25 diameters in height failing at knots and not by direct crushing of the fibres. Even small knots are injurious, but as these cannot be avoided in large scantlings, they must be accepted as an inherent defect of the material. Timber containing large and very numerous knots should, however, always be rejected.

Table V.[1]—BAUSCHINGER'S TESTS OF COMPRESSIVE STRENGTH OF RED DEAL (*Pinus silvestris*) AND WHITEWOOD (*Picea excelsa*)

Time of Felling.	Red Deal from Lichtenhof.		Whitewood from— Frankenhofen.		Whitewood from— Regenhütte.		Whitewood from— Schliersee.	
	1. lbs. per sq. in.	2. lbs. per sq. in.	1. lbs. per sq. in.	2. lbs. per sq. in.	1. lbs. per sq. in.	2. lbs. per sq. in.	1. lbs. per sq. in.	2. lbs. per sq. in.
Summer	5244	7183	4807	6416	5319	6287	3143	4570
Winter	6784	6343	5618	6756	5348	6343	4238	4779

Columns 1. Tested 3 months after felling. Columns 2. Tested 5 years after felling.

CHAPTER II

TENSILE AND SHEARING STRENGTH

1. TENSILE STRENGTH

Test specimens of good timber show that the material has a high tensile strength, but unfortunately it is impossible to utilize this strength to the full in actual construction. We have seen (fig. 514, p. 299) that the foot of a principal rafter tends to rupture the end of the tie-beam by shearing. In a similar manner, the end of a collar in a simple braced roof, as in fig. 566, may be subjected to shearing stresses from A to B and from the bolt C to D. And the nature of the material is such that it will yield to these shearing stresses long before the ultimate tensile strength of the tie-beam or collar is reached.

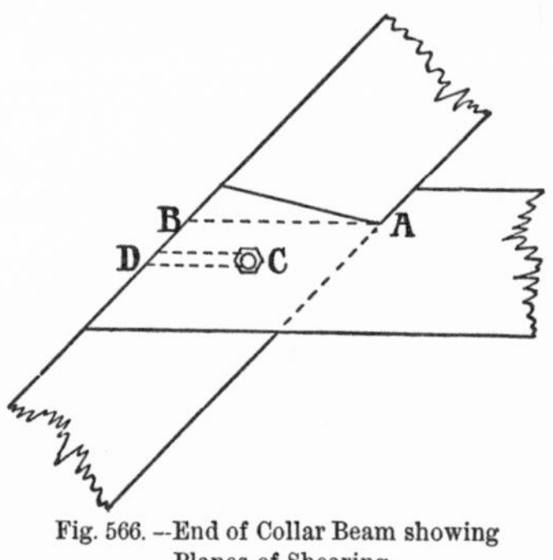

Fig. 566.—End of Collar Beam showing Planes of Shearing

The greatest difficulty experienced in ascertaining the tensile strength of wood is this liability of the material to shear. "In all cases", says Professor Lanza, "it has been found necessary to provide them [*i.e.* the test specimens] with shoulders, each shoulder being five or six times as long as the part of the specimen to be tested, and to bring upon these shoulders a powerful lateral pressure, to prevent the specimen from giving way by shearing along the grain, and pulling out from the shoulder instead of tearing apart."

In timber structures it is impossible to fulfil these conditions, and any calculations based on the sectional area and ultimate tensile strength of a wood tie are therefore bound to be erroneous. Moreover, it is only possible to test small specimens under tension, say, 1 inch square or less, and these give higher results than will be obtained with large scantlings in which knots and other defects invariably occur.

It may be stated, however, that experiments show that the ultimate strength of fir, spruce, and pines (except the pitch-pines) is about 8000 to 10,000 lbs. per square inch, of pitch-pine and English hardwoods about 13,000 to 15,000, and of the toughest Australian

[1] After Lanza.

woods (*Eucalyptus*) about 20,000. The ultimate tensile strength of wrought-iron is about 50,000 lbs. per square inch.

Some tensile tests of Canadian timber, recorded by Professor Bovey in a paper read before the Canadian Society of Engineers in 1895, may be summarized if only for the sake of showing the wide ranges of strength obtained. Fifty-one test-pieces of *Douglas Fir*, cut from nine different logs, yielded an average of 11,519 lbs. per square inch, the lowest result being 2485 lbs. and the highest 18,856 lbs.; eight tests of one log gave an average of only 6504 lbs., while the average of eight tests of another log was 15,734 lbs. Fourteen specimens of *Spruce*, cut from three different logs, gave an average of 9578 lbs. per square inch, but in this case the results were less erratic, the extremes being 7662 lbs. and 12,626 lbs. Six tests of *Red Pine* gave an average of 10,904 lbs. (maximum, 14,372 lbs.; minimum, 6274 lbs.). Only four tests of *White Pine* (sold in this country as *Yellow Pine*) were made, the range being from 10,347 to 12,969 lbs., and the average 11,887 lbs.

The modulus of tensile elasticity, although it has been given in many books on carpentry for different kinds of wood, is of little practical use. Specimens vary so much in their proportionate extension that it is impossible to obtain a modulus which shall be generally applicable. Moreover, the insuperable difficulties in the way of utilizing the tensile strength of timber in actual work would render such a modulus unnecessary, even if it could be found.

To ascertain the ultimate tensile strength of a piece of wood, it is merely necessary to multiply the area of the cross-section in square inches by the coefficient of tensile rupture for that kind of wood; thus—

$$\text{Ultimate tensile strength} = b\,d\,f \qquad (14)$$

Where b = breadth in inches, d = depth in inches, and f = the coefficient of tensile rupture.

2. SHEARING STRENGTH

The determination of the shearing strength of timber is of much more importance than that of the resistance to tension. We have already seen that a tensile stress in any member induces a compressive stress at right angles to the tensile stress; in the same manner, a shearing stress along one plane induces an equal similar stress along a plane at right angles to it. It follows, therefore, that calculations of shearing strength should be made on the basis of the smaller resistance, and as the longitudinal shearing resistance is the smaller, this ought to be the basis adopted. Some experiments carried out by Professor Bovey on Canadian timber are interesting in this connection. The tests were all made in the direction of the grain, but the planes of shearing were respectively (1) tangential to the annual rings, (2) radial, and (3) oblique. The following results were obtained:—*Douglas Fir*, (1) 411·61 lbs. per square inch (average of 38 tests), (2) 377·14 lbs. (29 tests), (3) 403·6 lbs. (30 tests); average of all tests, 398·82 lbs. *Red Pine*, (1) 392·77 lbs. (4 tests), (3) 363·68 lbs. (8 tests). *White Pine*, (1) 382·37 lbs. (4 tests), (2) 273 lbs. (2 tests), (3) 363·68 lbs. (3 tests). *Old Spruce*, (1) 332·29 lbs. (5 tests), (2) 389 lbs. (5 tests), (3) 383·13 lbs. (11 tests). In Douglas Fir and White Pine the radial plane is the weakest and the tangential the strongest, while in old Spruce the positions are reversed; but the tests of White Pine and Spruce were too few in number to be of much value.

The shearing strength of different species of wood varies chiefly according to the nature of the grain, and does not bear a uniform ratio to the compressive or transverse strength. A wood with contorted grain offers a greater shearing resistance, other things being equal, than one with a clean straight grain. American white oak has practically the same compressive strength as yellow pine, but has nearly 50 per cent more resistance to shearing.

In Table VI the resistance to shearing of the most important woods is stated according to various authorities. The results given in column No. 1 are those obtained by the U.S.A.

Table VI.—LONGITUDINAL SHEARING STRENGTH OF TIMBER

	Pounds per Square Inch.					
HARDWOODS	1.	2.	3.	4.	5.	6.
ACACIA, Blackwood (Australian, *Acacia melanoxylon*) ...	...	1537*d*	...	2350*b*	...	...
ALDER (*Alnus glutinosa*)	...	...	...	...	1464	...
ASH, English (*Fraxinus excelsior*)	...	...	...	1400*a*	1777	...
American White (*F. americana*)	1100	590*a*	...	...	...	...
BEECH (*Fagus silvatica*)	...	...	...	...	2118	...
BIRCH, Silver (*Betulus alba*)	...	...	...	...	1535	...
Yellow (*B. lutea*)	...	660*a*	...	...	...	...
CHESTNUT (*Castanea vulgaris*, var. *americana*)	...	...	690*a*	...	...	600
ELM, Cedar (*Ulmus crassifolia*)	1300	...	...	...	...	...
White (*U. americana*)	800	...	...	...	...	...
Common (*U. campestris*)	...	...	...	1400*a*	1905	...
EUCALYPTUS, Ironbark (*E. resinifera*)	...	1525*d*	...	2150*b*	...	...
Jarrah (*E. marginata*)	...	...	680*b*	1830*b*	...	...
Karri (*E. diversicolor*)	...	1154*c*	580*b*	...	...	...
Tuart (*E. gomphocephala*)	...	...	670*b*	...	...	...
GREENHEART (*Nectandra rodiœi*)	...	874*d*	...	...	...	...
GUM, Sweet (*Liquidambar styraciflua*). (See *Eucalyptus.*)	800	...	...	...	...	...
HICKORY, Pignut (*Hicoria glabra*)	1200	...	...	...	...	...
Shagbark (*H. ovata*)	1100	...	...	...	...	...
HORNBEAM (*Carpinus betulus*)	...	...	...	...	2118	...
LOCUST (*Robinia*)	...	...	1180*a*	...	...	...
MAPLE, White (*Acer saccharinum*)	...	500*a*	...	...	...	...
MORAWOOD (*Mora excelsa*)	...	1021*d*	...	...	...	...
OAK, American White (*Quercus alba*)	1000	840*a*	780*a*	...	...	800
„ Red (*Q. rubra*)	1100	810*a*	...	...	...	...
European (*Q. robur*)	...	...	...	2300*a*	1478	...
POPLAR, Aspen (*Populus tremula*)	...	...	...	...	1237	...
SOFTWOODS						
CEDAR, White (*Chamæcyparis thyoides*)	400	...	...	...	...	...
CYPRESS, Bald (*Taxodium distichum*)	500	...	...	...	...	...
DOUGLAS FIR (*Pseudotsuga douglasii*)	500	398*b*	...	...	...	...
FIR. (See *Pine*, *Spruce.*)						
HEMLOCK (*Tsuga*)	...	...	540*a*	...	...	350
LARCH (*Larix decidua*)	...	...	...	...	1336	...
PINE—Cuban (*Pinus cubensis* or *heterophylla*)	700	...	...	...	...	...
Loblolly (*P. tæda*)	700	...	...	...	...	...
Long-leaf (*P. palustris*)	700	370*a*	510*a*	...	...	600
Norway or Red (*P. resinosa*)	500	353*b*	...	...	...	400
Short-leaf (*P. echinata*)	700	...	...	...	...	400
Spruce Pine (*P. glabra*)	800	...	...	...	...	...
White Pine (*P. strobus*)	400	363*b*	490*a*	...	...	400
„ „ (Canadian)	...	...	...	...	...	350
Yellow or Red Deal or Fir (*P. silvestris*)	...	...	...	590*c*	1222	...
SEQUOIA or CALIFORNIAN REDWOOD (*Sequoia sempervirens*)	...	...	...	...	...	400
SPRUCE, American (*Picea alba* or *nigra*)	...	372*b*	470*a*	600*a*	...	400
Swedish Whitewood (*P. excelsa*)	...	...	...	...	1117	...

NOTES.—*Column No. 1* contains the results of tests by the U.S.A. Division of Forestry. *No. 2*, *a*, tests at the Watertown Arsenal, U.S.A.; *b*, tests of Canadian timber recorded by Professor Bovey; *c*, tests at the Belgian Government Testing Station, Malines; *d*, tests by Prof. Unwin. *No. 3*, *a*, those recorded by Hatfield as having been made on American woods; *b*, tests by Admiralty of Western Australia. *No. 4*, *a*, recorded by Rankine; *b*, tests made by W. H. Warren; and *c*, by Barlow. *No. 5* gives the results of a series of tests of Swedish woods recorded by Aug. Wijkander in his *Untersuchung der Festigkeits-Eigenschaften Schwedischer Holzarten* (Götëborg, 1897). *No. 6* contains the figures recommended by a committee of the "American Association of Railway Superintendents of Bridges and Buildings".

The low results obtained at the Watertown Arsenal (Col. No. 2, *a*) would seem to indicate some error in the method of testing.

The test of English oak recorded by Rankine (Col. No. 4) is very high, and cannot be considered a fair average.

Forestry Division. The test-pieces were 2 inches square and 8 inches long, and through each piece two holes, 1 inch square, were cut at a distance of 1 inch from each end, and at right angles to each other, as shown in fig. 567. A central hole was also bored to receive the pin of the testing apparatus. A metal key was passed through one of the square holes and fastened to a suitable stirrup, and when the shearing of this end of the specimen had been effected, the other end was similarly treated. Thus, tests on two planes at right angles to each other were obtained, and the mean of the two gave the shearing strength of the specimen. It will be observed that the shearing at each end was effected on two planes, measuring 2 inches by 1 inch, or a total area of 4 square inches. The mean results of the tests were therefore divided by 4 in order to obtain the shearing strength per square inch.

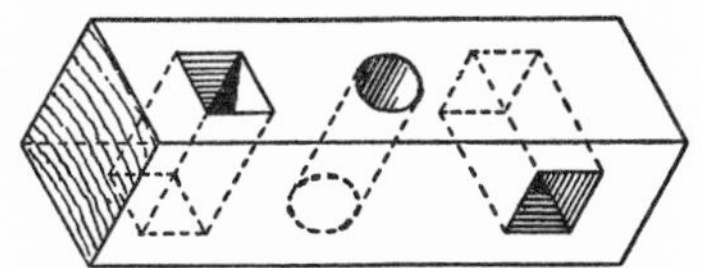
Fig. 567.—Test Specimens for ascertaining the Shearing Strength of Wood

To calculate the ultimate shearing resistance of a piece of wood, multiply the area (in inches) of the plane or planes along which rupture would occur, by the coefficient of rupture for the kind of wood, thus—

$$\text{Ultimate shearing strength} = af \qquad (15)$$

Where a = the area of the plane or planes in square inches, and f = the coefficient of shearing rupture.

Example 1.—Required the ultimate shearing strength of the end of a pitch-pine tie-beam 6 inches broad, into which a principal rafter is notched 9 inches from the end (fig. 514, page 299).

The area of the shearing plane will be 9 inches × 6 inches = 54 square inches, and the coefficient for pitch-pine (long-leaf) is stated to be 600 in column 6, Table VI. The ultimate shearing strength will therefore be 54 × 600 = 32,400 lbs.

Example 2.—Required the ultimate shearing strength of an oak king-post 7 inches square, into which collars are notched on both sides 8 inches from the end (fig. 568).

In this case there will be two planes of rupture, each measuring 8 inches by 7 inches. Therefore the ultimate shearing strength will be 2 (8 × 7) × (say) 1000 lbs. = 112,000 lbs.

In the case of a mortise at the end of a tie-beam there will be three planes of rupture, namely the planes in continuation of the two sides and of the bottom.

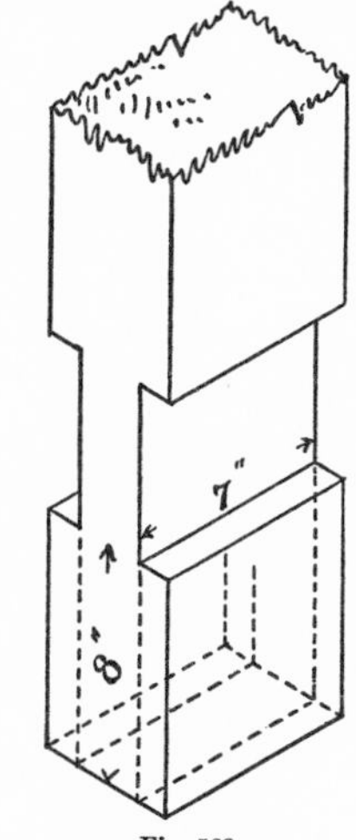

Fig. 568. Shearing Strength of Double-notched King-post

Shearing across the fibres may occur in certain cases, as in trenails. Experiments by Parsons on English oak trenails gave an ultimate strength of about 4000 lbs. per square inch. The values recommended by the Committee of the "American Association of Railway Superintendents" for shearing "across grain" are as follows:—Chestnut, 1500 lbs. per square inch; white oak, 4000; long-leaf yellow pine, 5000; short-leaf yellow pine, 4000; spruce, 3000; hemlock, 2500; white pine, 2000; cedar, 1500.

CHAPTER III

COMPRESSIVE STRENGTH

The behaviour of timber under compression is much more satisfactory than under tension. Not only is testing easier, but the conditions under which the tests can be made do not differ so widely from the conditions prevailing in practice as to render the tests of little value.

Tests made on small specimens are, however, unsatisfactory for the reasons already given, and we will therefore confine ourselves, as much as possible, to tests of large scantlings. Among the earliest of these were the tests made at the end of the eighteenth century by a Frenchman named Girard, the results being published in 1798. More recently large scantlings have been tested in this and other countries, but there are still many kinds of wood of which the compressive strength has not yet been properly ascertained. A comprehensive inquiry into the strength of timber was instituted in 1891 by the Division of Forestry of the United States Department of Agriculture, and extremely valuable results were obtained, but unfortunately the enquiry has never been completed, and Mr. J. W. Toumey, the Acting Forester, writing on April 25, 1900, said: "There is no prospect of further work of this character being done".

The earliest carpenters can scarcely have failed to discover the fact that long posts are weaker than short posts of the same scantling, but Rondelet, who published his *Traité Théorique et Pratique de l'Art de Bâtir* in 1805–1810, appears to have been the first to undertake an elaborate series of tests for the purpose of ascertaining the relative strength of posts of different heights. His experiments led him to formulate the following ratios[1] for the strength of square posts, the strength of a cube being taken as the unit, and the height being expressed in terms of a side of the square cross-section:—

Table VII.—The Strength of Square Wood Posts according to Rondelet

Length in Terms of the Breadth.	Proportionate Strength.
1	1
12	$\frac{5}{6}$
24	$\frac{1}{2}$
36	$\frac{1}{3}$
48	$\frac{1}{6}$
60	$\frac{1}{12}$
72	$\frac{1}{24}$

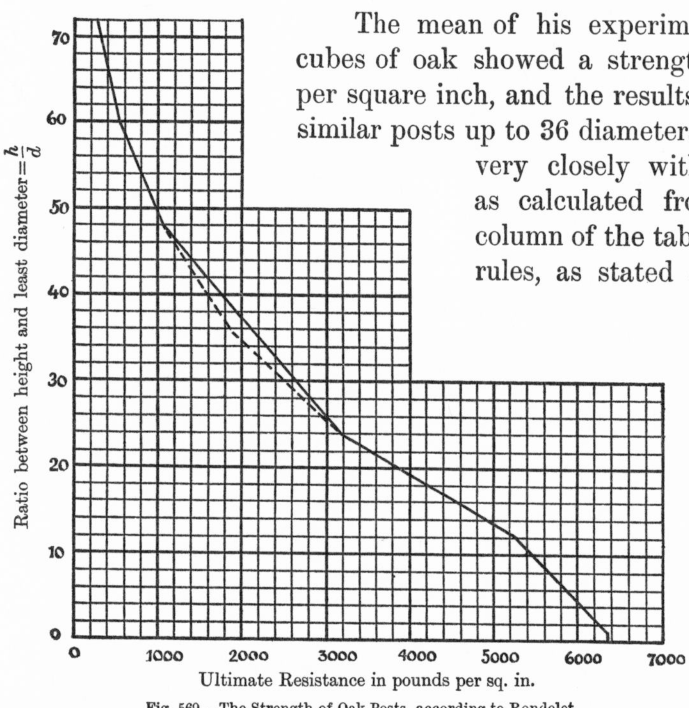

Fig. 569.—The Strength of Oak Posts, according to Rondelet

The mean of his experiments on 1-inch cubes of oak showed a strength of 6346 lbs. per square inch, and the results of his tests of similar posts up to 36 diameters high accorded very closely with the strength as calculated from the second column of the table. Rondelet's rules, as stated above, are unsatisfactory in so far as they deal only with certain specified ratios of length to breadth, but this objection can be overcome by expressing the strength graphically, as in fig. 569, where the irregular diagonal line represents the ultimate strength of oak posts per square inch, the strength of a 1-inch cube being taken at 6346 lbs. In the diagram each horizontal division represents a strength of 200 lbs. per square inch, while the vertical divisions represent the height in relation to the least diameter. It is clear from the diagram that a more regular curve would be obtained by making the strength of posts 36 diameters high $\frac{7}{24}$ of the maximum strength, as shown by the dotted lines, instead of $\frac{1}{3}$ as given in the table, and shown by the full line.

Many attempts have been made to devise a formula for calculating the strength of timber posts of various heights, among the best-known being that of Hodgkinson (1840), which is based on the assumption that the ultimate strength of long round or square pillars varies directly as the fourth power of the shortest diameter, and inversely as the square of the length; that is to say—

$$w \propto \frac{d^4}{l^2},$$

where w = breaking weight or ultimate strength, d = shortest diameter, and l = length.

[1] See Gwilt's *Encyclopædia of Architecture*, p. 416, 1888 edition.

This theory is really due to Euler, who had come to the conclusion that relatively-short pillars give way by crushing, while relatively-long pillars give way by bending. By the terms "relatively-short" and "relatively-long" are meant pillars whose length in proportion to their diameter is below or above a certain ratio; this ratio Euler assumed to vary for different materials. According to him, therefore, the strength of "short" pillars is simply the area of the cross-section in square inches multiplied by the weight required to crush an inch cube of the same material; that is to say—

For wood pillars less than 30 diameters in height, $W = f\,b\,t$, ... (16)

where W = ultimate crushing weight, f = crushing weight of an inch cube, b = breadth in inches, and t = thickness in inches.

For "long" pillars he proposed a formula which included the modulus of elasticity of the material, and other factors, among which the most important are those referred to above, namely, that the strength varies as $\frac{d^4}{l^2}$.

Hodgkinson adopted Euler's main assumptions, but simplified the formula for long pillars and altered it to accord more nearly with the results of tests which he made on the compressive strength of wood. His formula for rectangular pillars more than 30 diameters high, with flat ends, is—

$$W = C\,\frac{b \times t^3}{L^2} \qquad (17)$$

where W = ultimate crushing weight in tons, b = breadth (or longer side) of the pillar in inches, t = thickness (or shorter side) of the pillar in inches, L = length of the pillar in *feet*, and C = a constant, which he calculated to be 10·95 for dry oak, and 7·81 for dry fir.

Another well-known formula is that introduced into this country by Gordon and usually known as Rankine's; it is based on the assumption that all pillars yield under a combination of crushing and bending strains. The formula for pillars with flat ends is as follows:—

$$W = \frac{f A}{1 + \frac{l^2}{C r^2}} \qquad (18)$$

where W = breaking weight in lbs., f = crushing strength of a cube of the material in lbs. per square inch, A = area of the cross-section in square inches, C = a constant, l = length of the pillar in inches, and r = radius of gyration. In rectangular pillars $r^2 = \frac{t^2}{12}$, t being the thickness or shorter side of the cross-section in inches; in circular pillars $r^2 = \frac{d^2}{16}$, d being the diameter in inches. For "timber", Rankine, following Weisbach, makes $f = 7200$, and $C = 3000$ if used with r^2, or 250 if used with t^2. The formula may therefore be expressed thus, for rectangular pillars—

$$W = \frac{7200\,A}{1 + \frac{l^2}{250\,t^2}} \qquad (19)$$

This is the formula adopted in the Chicago building regulations, but with different values for f; instead of 7200 a coefficient of 600 is employed for "white or Norway pine", 800 for "oak", and 900 for "long-leaf yellow pine". These values give the *safe* loads which pillars are calculated to bear, and not the breaking weights. Mr. C. Shaler Smith, an American writer who has carried out a great number of tests on pillars, adopts the same formula, but makes $f = 5000$ for well-seasoned "yellow pine", the result giving the breaking weight.

Careful comparison of these formulas shows how unsatisfactory and discrepant they are, and how much the strength calculated from them differs from the strength obtained by actual tests. In fig. 570 the curve of strength (H–H) for oak pillars as calculated from Hodgkinson's formula is shown, and also the actual tests of small oak sticks on which his formula was based; Rankine's curve (R–R) for "timber" is also drawn, and other actual tests of oak, together with the curve (L S) calculated from the author's formula, are plotted for purposes of comparison. Fig. 571 gives the actual and calculated strength of red deal pillars in a similar manner.

It will be observed that Hodgkinson's formula accords pretty closely *with his own experiments* within certain limits of length, both for oak and red deal, but that it does not

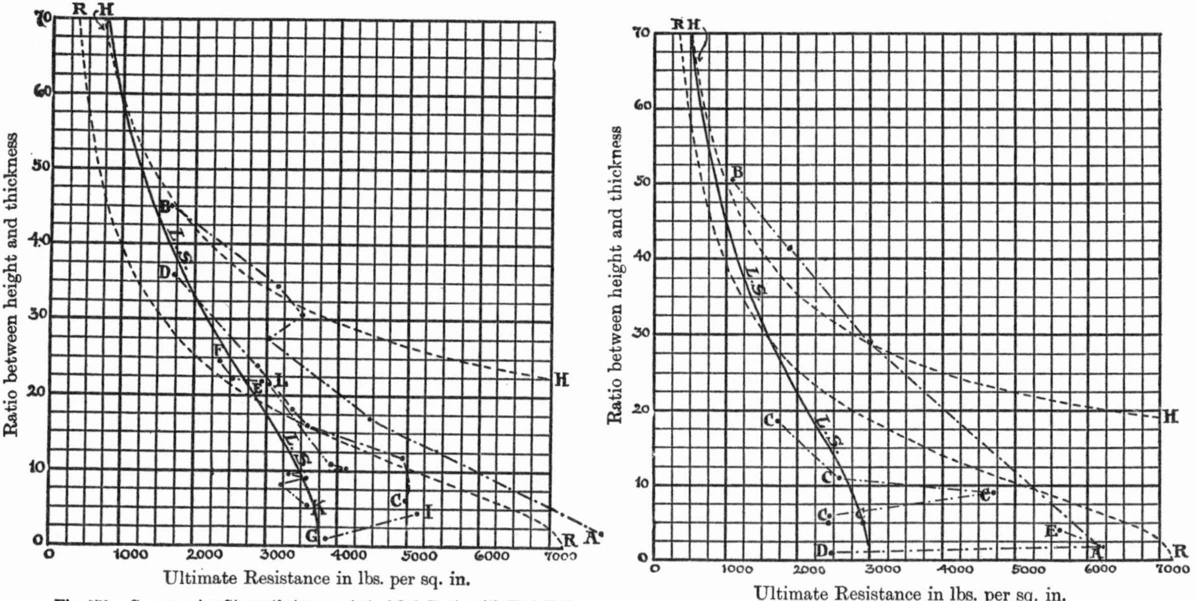

Fig. 570.—Compressive Strength (per sq. in.) of Oak Posts with Flat Ends

A–B, Hodgkinson's tests of Dantzic oak sticks from 1 to $1\frac{3}{4}$ in. square; C–D, Lamandé's tests of French oak posts from $2\frac{1}{8}$ to $4\frac{1}{4}$ ins. square; E–F, Girard's tests of French oak posts about 6 ins. by 4 ins.; H–H, strength of "dry oak" according to Hodgkinson's formula (17); G–I, Rennie's tests of English oak 1 in. square; K–L, Kirkaldy's tests of English oak (large scantlings); R–R, strength of "timber" according to Rankine's formula (19); LS, strength of oak according to the author's formula (23).

Fig. 571.—The Strength (per sq. in.) of Red Deal Posts with Flat Ends

A–B, Hodgkinson's tests of pillars from 2 ins. by 2 ins. to $3\frac{1}{2}$ ins. by $1\frac{1}{4}$ in.; CCCCC, Kirkaldy's tests of pillars from 5 ins. by 5 ins. to about $13\frac{1}{2}$ ins. by $12\frac{3}{4}$ ins.; D–E, tests by the Royal Inst. of Brit. Arch. of pillars from 3 to 4 ins. square; H–H, strength of red deal according to Hodgkinson's formula (17); R–R, strength of "timber" according to Rankine's formula (19); L S, strength of red deal according to the author's formula.

accord with Lamandé's and Girard's tests of oak, and at the best is only applicable to pillars above 30 diameters high for red deal and 35 for Dantzic oak. Rankine's theoretical strength of "timber", as applied to oak, is less than half the actual strength found by Hodgkinson for pillars between 30 and 45 diameters high, but accords very closely with Lamandé's and Girard's tests of larger scantlings of French oak from about 11 diameters high upwards. Applied to red deal (fig. 571), it is still very much below the results obtained by Hodgkinson for pillars between 29 and 50 diameters high, but is much higher than the strength (marked C) obtained by Kirkaldy on large red deal posts, and for short pillars generally it gives a strength far in excess of actual tests.

The general conclusions from these comparisons are that Hodgkinson's formula and constants give results much too high, and that at the best the formula is applicable only within very restricted limits. Rankine's formula *appears* to be practically accurate for oak posts between 10 and 25 diameters in height, but a consideration of it in comparison with more recent tests will show that it requires alteration not only for different varieties of timber, but also in other respects.

The most complete tests for our purpose are those carried out on posts of "white pine" and "yellow pine" at the Watertown Arsenal, U.S.A., and plotted in figs. 572 to 574. The "white pine" referred to is the wood usually termed in this country "yellow pine", and the "yellow pine" is chiefly known to us, though incorrectly, as "pitch pine". In this article the American names will be used.

The tests of "white pine" are shown in fig. 572. Each point plotted on the diagram is the average of three tests, with the exception of A (one test), the next point above A (two tests), and the point midway between E and F (six tests).

The "yellow pine" tests are shown in figs. 573 and 574. Those from A to B (fig. 573) give the average results obtained on posts about $5\frac{1}{2}$ inches square. The actual results of all the tests in this series are plotted in fig. 574, in order that the great difference between the highest and lowest tests can be seen at a glance; of the three posts 33 diameters high, the weakest bore only about three-fourths as much as the strongest, while at $62\frac{1}{2}$ diameters high the weakest was only two-thirds the strength of the strongest.

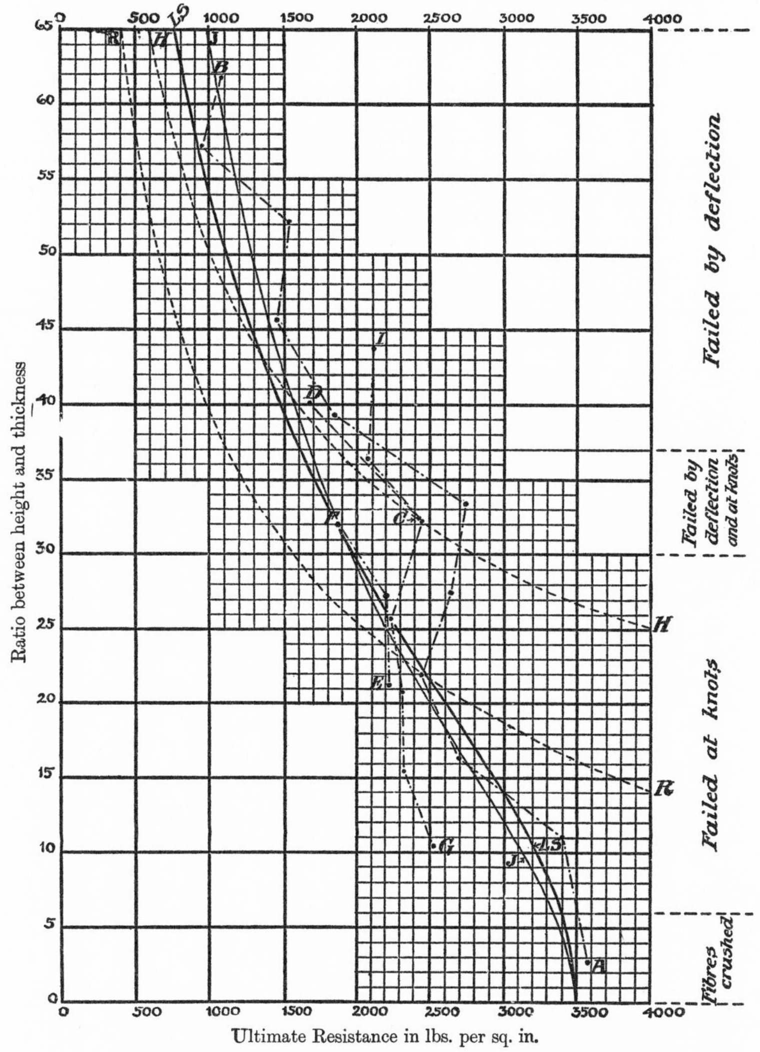

Fig. 572.—The Strength of American "White Pine" Pillars with Flat Ends

A–B, Posts from $5\frac{1}{8}$ to $5\frac{1}{2}$ ins. square; C–D, posts $5\frac{1}{2}$ ins. × $11\frac{5}{8}$ ins., and $4\frac{1}{2}$ ins. × $11\frac{5}{8}$ ins.; E–F, posts from $5\frac{5}{8}$ ins. × $15\frac{5}{8}$ ins. to $8\frac{1}{2}$ ins. × $16\frac{1}{2}$ ins.; G–I, posts $7\frac{5}{8}$ ins. × $9\frac{5}{8}$ ins.; H–H, strength of "red deal" posts according to Hodgkinson's formula; R–R, strength of "timber" posts according to Rankine's formula; J–J, strength according to Johnson's formula but with different coefficient; LS, strength according to the author's formula.

To return to fig. 573; the line from C to D records the tests on posts about $7\frac{3}{4}$ inches by $9\frac{3}{4}$ inches, each point except one being the average of three tests; E to F gives similar averages for posts about $15\frac{3}{4}$ inches wide, and from $5\frac{5}{8}$ to $8\frac{1}{4}$ inches thick; and G to I shows the results of single tests of posts from $4\frac{1}{2}$ inches by $11\frac{5}{8}$ inches to $5\frac{7}{8}$ inches by $11\frac{3}{4}$ inches.

It was found that very short pillars, up to about 5 diameters high, failed usually by the crushing of the fibres; pillars between 5 and about 25 or 30 diameters high failed at knots; slightly longer pillars failed either at knots or by deflection, and pillars above about 35 diameters high failed wholly by deflection.

Knots are the chief source of weakness in pillars less than about 25 or 30 diameters high. The wide divergence of the tests of these short pillars and the general lack of increase of strength below this limit are almost wholly due to knots.

It is clear that Rankine's formula is extremely inaccurate. Applied to yellow pine (fig. 573), it gives too great a strength for short posts, and much too small a strength for posts above 15 or 20 diameters in height; between 25 and 35 diameters high the calculated breaking weight is less than half the actual, and above 45 or 50 diameters it is less than one-third. As Rankine recommends a factor of safety of 10, it follows that long posts, designed in accordance with his formula, would be loaded with only one-twentieth or one-thirtieth of the actual breaking weight. This involves an unnecessary waste of material. Hodgkinson's formula, with proper constants for different kinds of timber, will give fairly accurate results, as shown in figs. 572 and 573, for posts above 30 or 35 diameters high. As, however, it is not of general application, we need not consider it further.

The formula proposed by C. Shaler Smith, and shown by the line S S–S S in fig. 573, is simply Rankine's formula with a lower coefficient for the purpose of making it applicable to short pillars. The result is, that the strength of long pillars is still more grossly under-estimated than by Rankine's formula.

Three additional formulas have been proposed by American writers in recent years. They are Merriman's, Johnson's, and Stanwood's. Of these, Stanwood's is the simplest. It

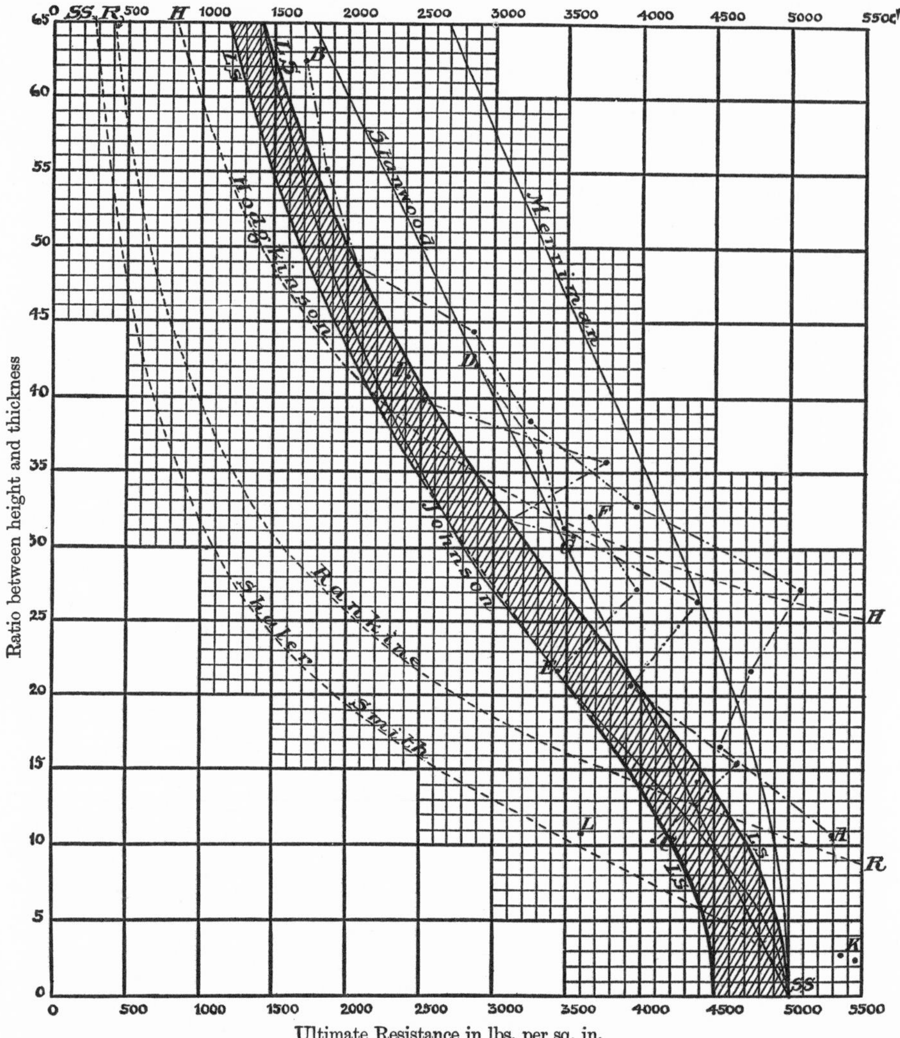

Fig. 573.—The Strength of American "Yellow Pine" Pillars with Flat Ends

A–B, Posts about $5\frac{1}{2}$ ins. square; C–D, posts about $7\frac{5}{8}$ ins. × $9\frac{3}{4}$ ins.; E–F, posts from $5\frac{5}{8}$ ins. × $15\frac{1}{2}$ ins. to $8\frac{1}{4}$ ins. × $16\frac{1}{4}$ ins.; G–I, posts from $4\frac{1}{2}$ ins. × $11\frac{5}{8}$ ins. to $5\frac{7}{8}$ ins. × $11\frac{3}{4}$ ins.; K, posts 9 ins. square and 10 ins. × $10\frac{1}{8}$ ins.; L, post about $5\frac{1}{2}$ ins. square; and the curves of strength calculated according to the formulas of Rankine, Hodgkinson, Shaler Smith, Merriman, Stanwood, Johnson, and the author (the latter, marked L S–L S, is plotted in duplicate, the value of f being 4500 in the one case and 5000 in the other).

is designed to give the *safe* load (with a factor of safety of 5) and not the breaking weight, and may be expressed thus—

$$\text{Safe load} = c - 10\frac{l}{d}, \quad \ldots\ldots \quad (20)$$

where c = the coefficient (1000 for "yellow pine"), l = length in inches, and d = the shorter side of the cross-section in inches.

This formula is expressed graphically by a straight line and not by a curve, and differs therefore from all the other formulas which have been, or will be, considered. For the purpose of comparing it with the other formulas and with the results of the tests on "yellow pine" posts, the calculated safe loads have been multiplied by 5 (the factor of safety) in order to obtain the breaking weight, and the line of strength has been plotted in fig. 573. It will be observed that this accords remarkably well with the actual tests for pillars up to about 40 diameters high, but above this limit the formula gives too high a strength; thus, according to the formula the safe load for a post 50 diameters high should be 2500 lbs. per square inch, but the actual tests gave an average breaking strength of 2025 and a minimum strength of only 1820 lbs., so that the factor of safety allowed by the formula would be only 4 for the average and $3\frac{2}{3}$ for the minimum. This is certainly a lower factor than ought to be allowed in timber structures. The formula, however, has the merit of simplicity, but in using it a deduction of, say, one-fifth ought to be made from the calculated safe load in order to give a sufficient margin of safety.

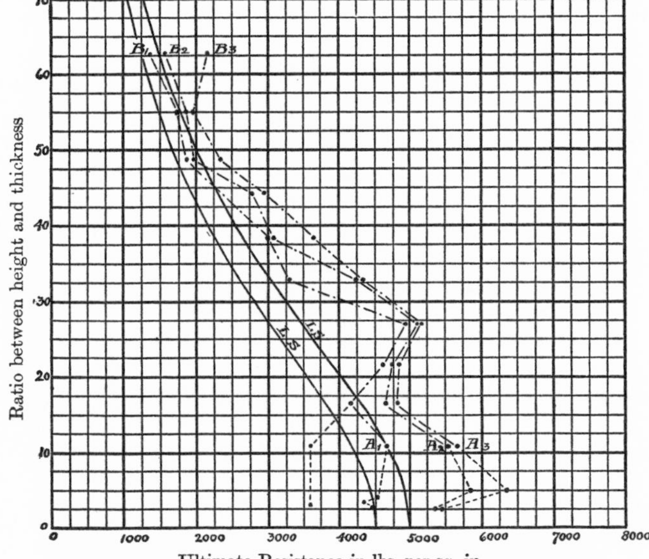

Fig. 574.—Single Tests of American "Yellow Pine" Pillars with Flat Ends

Note.—The tests from A_1 to B_1, A_2 to B_2, and A_3 to B_3 give the actual breaking weights of the pillars, the mean strength of which is shown by the line A–B in fig. 573; the remaining points are single tests of short pieces.

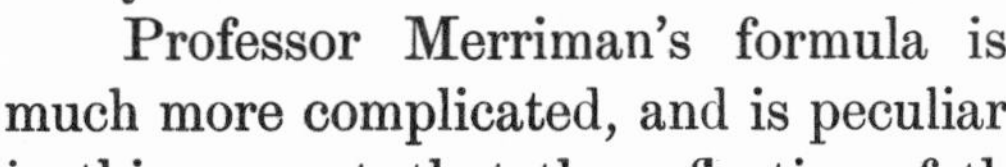

Professor Merriman's formula is much more complicated, and is peculiar in this respect, that the reflection of the curve of strength for "yellow pine", in the graphic representation of the formula, does not commence till a height of 45 diameters has been reached. For "yellow pine" the formula is—

$$\text{Safe load} = \frac{1000}{1 + \frac{1}{5000}\left(\frac{l}{d}\right)^2} \quad \ldots\ldots \quad (21)$$

If we multiply the safe load obtained by the formula by a factor of safety of 5, we obtain the curve of breaking strength plotted in fig. 573. The formula is dangerously unsafe for all posts above 30 diameters high.

Professor A. L. Johnson's formula looks more complicated than it really is. It is as follows:—

$$\text{Strength} = \frac{700 + 15\frac{l}{d}}{700 + 15\frac{l}{d} + \left(\frac{l}{d}\right)^2} \times \text{C}, \quad \ldots\ldots \quad (22)$$

where l = length in inches, d = least diameter in inches, and C a constant varying according to the species of wood. The value of C for yellow pine is 1000, if the safe load is required, or 5000 for the breaking strength.

This formula appears to me to be the best of those which have yet been considered. It agrees remarkably well with the actual tests of a weak wood such as white pine[1] (see the curve J–J, fig. 572), and a strong wood such as yellow pine (fig. 573); in the case of yellow pine, for example, it approximates fairly closely to the tests throughout the range of height, although it gives perhaps too low a strength for posts between 20 and 50 diameters high; it errs in the other direction for white pine posts above 55 diameters in height.

Before my acquaintance with this formula of Professor Johnson's, I had devised a formula which gives a curve of strength in very close accord with the actual tests repre-

[1] The value of C for this wood is assumed to be 3500.

sented graphically in the foregoing illustrations. The curve is not much different from that obtained from Professor Johnson's formula, but it has the advantage of being somewhat safer for very long white pine columns. It is applicable to woods of varying strengths, as shown by the lines L S–L S in figs. 570 to 576, and has been adopted as the basis of calculation for the graphic representation of the strength of columns in Plate XXVI.

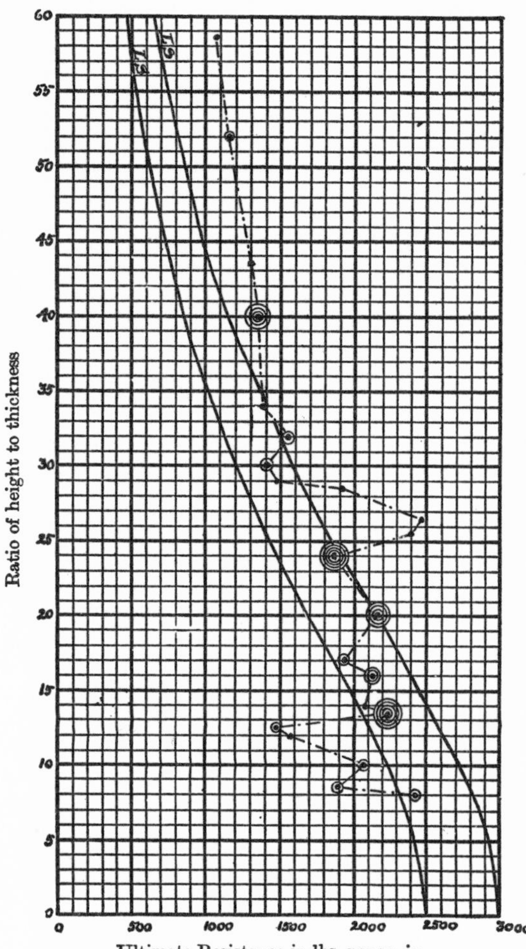

Fig. 575.—Tests of American "White Pine" Posts, carried out for Brunel in 1846

L S, Curves of strength according to the author's formula.

We have seen that Gordon's (Rankine's) formula includes both the coefficient of rupture and a constant, the latter remaining the same for different kinds of wood. Rankine's coefficient was too high, but a modification of the constant was also found to be necessary, and it was further seen that the same constant could not be satisfactorily used for woods of different strengths; in other words, the coefficient and constant must vary for different woods. To avoid the confusion which would have arisen from the use of a varying "constant", I tried to substitute for it an expression in terms of the coefficient. A satisfactory solution appears to have been found by adopting $\frac{f}{3}$ in lieu of the constant; we thus get an expression varying according to the strength of the wood in place of the unvarying constant of Rankine's formula. The new formula for rectangular pillars thus becomes—

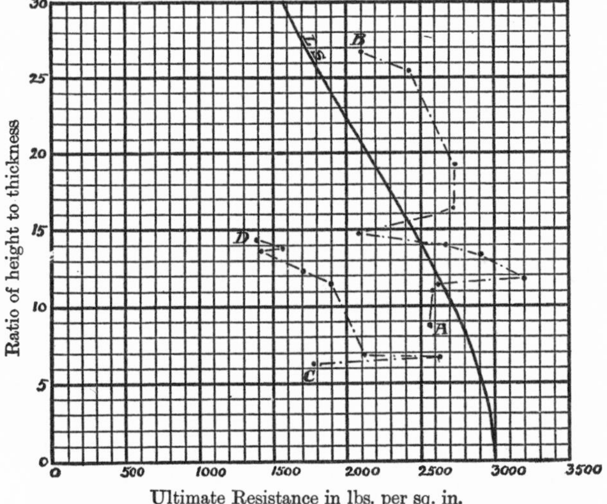

Fig. 576.—Tests of American Spruce Posts, carried out at the Massachusetts Institute of Technology

A–B, Posts from about 7¾ ins. square to 7¾ ins. by 12 ins.; C–D, posts tested with wood bolster at one end; L S, curve of strength according to the author's formula.

$$W = \frac{f A}{1 + \frac{3 r^2}{f}} \quad \text{.......... (23)}$$

where W = crushing weight in lbs. per square inch, A = area of cross section in square inches, r = ratio of length to smaller side of cross-section, that is, $\frac{\text{length}}{\text{thickness}}$, and f = coefficient of rupture varying for different kinds of timber.

Some tests of white pine carried out for Brunel in 1846, and recorded in Seddon's *Builders' Work and the Building Trades*, give a smaller strength than that obtained by the Watertown tests. Brunel's tests (fig. 575) are very erratic; this and the low breaking

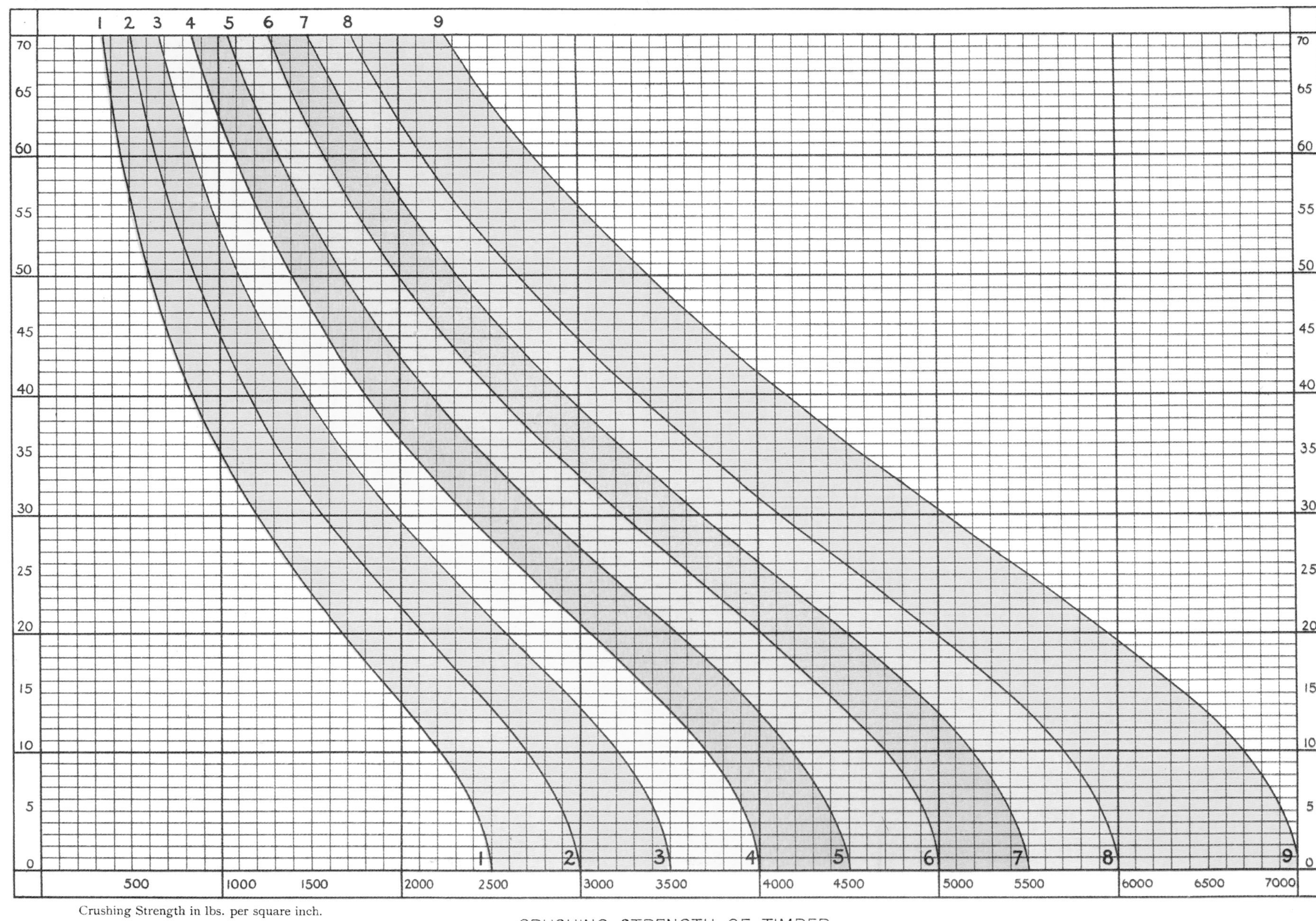

CRUSHING STRENGTH OF TIMBER

1–2. Softwoods: unseasoned or of inferior quality. 2–3. White (yellow) Pine, Spruce, White Deal, Yellow (red) Deal or Fir, Larch. 3–4. Cedar, Sequoia, Red Pine, Douglas Fir, Kauri Pine, Cypress. 4–5. Shortleaf Yellow Pine, Oak, Elm. 5–6. Yellow (pitch) Pine, Mahogany, Chestnut, Teak. 6–7. Hickory. 7–8. Jarrah, Karri, Blackwood. 8–9. Ironbark, Greenheart, Morawood.

weights may perhaps be due to the posts being tested with wooden bolsters at the ends; but this is only a supposition, as Colonel Seddon says nothing as to the method of testing. Nothing, moreover, is said as to the quality and degree of moisture of the timber. The wood used by Brunel is described as "American yellow pine (specified to be of Quebec or upper port growths)"; it is therefore the wood known in this country as "yellow" pine and in America as "white". The scantlings ranged from 8 to 15 inches square. In the diagram, a dot represents a single test, and each circle an additional test; where more than one test was made, the average is plotted.

Some tests of spruce columns carried out at the Massachusetts Institute of Technology are plotted in fig. 576. The tests from A to B refer to posts from about $7\frac{3}{4}$ inches square to $7\frac{7}{8}$ inches by 12 inches. The tests from C to D show the weakening effects of wood bolsters; these posts were tested with a bolster of oak, maple, or yellow pine at one end, and in each case the post split at the end next the bolster; the bolsters reduced the strength of the posts about one-third.

The values of the coefficient to be adopted for all the different species of wood now in use have not yet been accurately determined. The following table gives some of the most trustworthy results now available, together with the coefficients recommended by the committee of the American Association of Railway Superintendents:—

Table VIII.—COMPRESSIVE STRENGTH OF TIMBER IN POUNDS PER SQUARE INCH

1. Tests of small cubes; L. = Laslett. The figures in italics are the averages of tests of 1-, 2-, 3-, and 4-inch cubes; the others, of 2-inch cubes only, except Karri, of which a 6-inch cube was tested.
2. Tests of "short" columns; K. = Kirkaldy; W. = Professor Warren (Sydney University), tests of thoroughly-seasoned Australian timbers 3 ins. by 3 ins.; R. = Royal Inst. of Brit. Architects.
3. S. = tests of Swedish timber, recorded by Aug. Wijkander, in his *Untersuchung der Festigkeits-Eigenschaften Schwedischer Holzarten* (1897); A. = tests of American timber, carried out at the Watertown Arsenal, U.S.A.; F. = timber from Millar's Forests, tested by W. H. Stanger; M. = tests carried out at the Massachusetts Inst. of Technology.
4. A. = tests by the Admiralty of Western Australia; B. = tests of Canadian timber recorded by Prof. Bovey; U. = tests of Colonial timbers by Prof. Unwin (chiefly 3 ins. sq. and 8 ins. high).
5. Tests of green timber carried out by the U. S. A. Division of Forestry (Circular No. 15).
6. Do. do. corrected to 15 per cent moisture for Cuban, Loblolly, Long-leaf, and Short-leaf Pines, and 12 per cent for other woods.
7. Recommendations of a committee of the American Association of Railway Superintendents of Bridges and Buildings, for "columns under 15 diameters".

Description of Timber.	1.	2.	3.	4.	5.	6.	7.
HARDWOODS.							
ACACIA, Blackwood (*Acacia melanoxylon*) ..	...	7,184, W.	...	7,272, U.	...	...	...
ALDER (*Alnus glutinosa*)	...	...	5375, S.	...	...	...	...
ASH, American White (*Fraxinus americana*) ...	5,494, L.	...	...	...	...	7,200	...
Common (*Fraxinus excelsior*)	6,964, L.	3,025, K.	6086, S.	...	...	...	...
BAYWOOD. (See *Mahogany*.)							
BEECH (*Fagus silvatica*)...	...	...	6854, S.	...	...	...	...
BIRCH, Silver (*Betulus alba*)	...	...	6128, S.	...	...	...	...
BULLET-WOOD (*Mimusops kauki*)	...	...	...	10,687, U.	...	...	...
CHESTNUT, American (*Castanea vulgaris*, var. *americana*)	...	...	...	...	...	...	5000
CRABWOOD (*Carapa guianensis*)	...	...	...	7,385, U.	...	...	...
ELM, Cedar (*Ulmus crassifolia*)...	...	...	...	...	...	8,000	...
Common (*U. campestris*)	5,785, L.	...	6128, S.	...	...	...	...
White (*U. americana*)	*8,584*, L.	...	...	...	...	6,500	...
EUCALYPTUS, Blue Gum (*Eucalyptus* ——) ...	6,894, L.	...	...	...	...	...	...
Ironbark (*E. resinifera*)... ...	10,306, L.	11,000, W.	...	8,971, U.	...	...	...
Jarrah (*E. marginata*)	7,163, L.	6,800, W.	8948, F.	6,900, A.	...	...	...
Karri (*E. diversicolor*)	11,513, L.	6,560, W.	6837, F.	6,800, A.	...	...	...
Tallow-wood, and Red, Grey, and Flooded Gum ...	...	8,500, W.	...	9,000, U.	...	...	...
Tuart (*E. gomphocephala*) ...	...	...	...	9,300, A.	...	...	...
GREENHEART, Demerara (*Nectandra rodiæi*) ...	*14,420*, L.	9,629, K.	...	9,825, U.	...	...	...
GUM, Sweet (*Liquidambar styraciflua*) ... ... (See *Eucalyptus*.)	...	...	...	...	3300	7,100	...

Description of Timber.	1.	2.	3.	4.	5.	6.	7.
Hardwoods—*Continued.*							
Hickory, Pecan (*Hicoria pecan*)	...	...	...	...	3600	9,100	...
Pignut (*H. glabra*)	...	...	...	...	5400	10,900	...
Shagbark (*H. ovata*)	...	...	...	...	5700	9,500	...
Hornbeam (*Carpinus betulus*)	8,312, L.	...	6498, S.	...	...	...	...
Ironbark, Jarrah, and Karri. (See *Eucalyptus.*)							
Lignum-Vitæ (*Guaiacum officinale*)	...	...	...	8,672, U.	...	...	...
Lime (*Tilia parvifolia*)	...	...	4578, S.	...	...	...	...
Mahogany, African (*Khaya senegalensis*) ...	*10,002*, L.	...	...	...	...	...	...
Baywood (*Swietenia mahagoni*) ...	*6,395*, L.	6,373, K.	...	...	...	...	...
Spanish (*S. mahagoni*)	*6,413*, L.	...	...	...	...	...	...
Morawood (*Mora excelsa*)	...	...	...	9,498, U.	...	...	...
Oak, American White (*Quercus alba*)	*6,067*, L.	3,351, K.	3506, A.	...	5300	8,500	4500
American Red (*Q. rubra*)	...	...	...	...	...	7,200	...
European (*Q. robur*)	*7,475*,[1] L.	3,607, K.	5958, S.	...	...	...	...
Poplar, Aspen (*Populus tremula*)	...	...	5062, S.	...	...	...	...
Sabicu	*8,458*, L.	...	...	...	...	...	...
Satinwood from Jamaica	...	...	...	9,669, U.	...	...	...
„ „ Ceylon	...	...	...	7,558, U.	...	...	...
Teak (*Tectonis grandis*)...	*5,732*, L.	6,200, R.	...	...	...	...	...
Softwoods.							
Californian Redwood. (See *Sequoia.*)							
Cedar, Himalayan, or Deodar (*Cedrus deodara*)	...	4,265, K.	...	...	...	...	...
White (*Chamæcyparis thyoides*) ...	...	...	...	...	2900	5,200	...
Variety not stated	4,480, L.	...	...	...	...	...	4000
Cypress, Bald (*Taxodium distichum*)	...	...	...	...	4200	6,000	4000
Douglas Fir or Spruce (*Pseudotsuga douglasii*)	...	6,586, K.	...	6,120, B.	...	5,700	...
Fir. (See *Pine.*)							
Hemlock (*Tsuga*)	...	...	...	...	...	...	4000
Kauri Pine (*Agathis australis*)	*6,420*, L.	...	...	4,562, U.	...	...	...
Larch (*Larix decidua*)	*5,815*, L.	...	5076, S.	...	...	...	...
Oregon Pine. (See *Douglas Fir.*)							
Pine, Cuban (*Pinus heterophylla* or *cubensis*) ...	...	...	...	...	4800	7,850	...
Deodar. (See *Cedar.*)							
Loblolly (*Pinus tæda*)	...	...	...	...	4100	6,500	...
Long-leaf (*P. palustris*)	...	...	...	...	4300	6,850	...
Pitch. (See *Cuban, Loblolly, Long-leaf, Short-leaf,* and *Yellow.*)							
„ variety not stated...	6,462, L.	4,588, K.	...	...	...	...	...
Red or Norway (*Pinus resinosa*) ...	*5,684*, L.	...	...	4,067, B.	...	6,700	4000
Short-leaf (*P. echinata*)	...	...	...	...	3300	5,900	4000
Spruce (*P. glabra*)	...	...	...	...	3900	7,300	...
White[2] (*P. strobus*)	...	3,934, K.	3570, A.	...	...	5,400	3500
„ (Canadian)	*4,205*, L.	...	...	3,308, B.	...	...	5000
Yellow. (See above, *Cuban, Loblolly, Long-leaf,* and *Short-leaf.*)	...	...	4658, A.	...	...	...	5000
Yellow or Red Deal or Fir (*Pinus silvestris*)	*6,097*, L.	2,914,[3] K.	5887, S.	...	...	...	...
Sequoia (*Sequoia sempervirens*)...	...	...	...	...	...	...	4000
Spruce, American (*Picea alba* or *nigra*) ...	4,851, L.	...	2391, M.	4,377, B.	...	...	4000
Whitewood or White Deal (*Picea excelsa*)	...	5,376,[4] K.	5247, S.	...	...	...	...

The figures recorded by the U.S.A. Department of Agriculture are probably the most trustworthy, as they represent the averages of a large number of tests, and take into consideration the amount of moisture in the timbers; column 5 gives the actual crushing strength of green wood containing about 40 per cent of moisture, and column 6 gives the estimated values for thoroughly-seasoned wood. It must, however, be pointed out that, in the case of yellow pine, the averages appear to include tests of posts up to 16 diameters in height; these tests, according to the table in Circular No. 12, gave on the average nearly 4 per cent less strength than small test-pieces cut from the same trees.

The value to be attached to the results recorded in column 6 of Table VIII will

[1] Seasoned English oak; the strength of unseasoned English oak was 4920 lbs. per sq. in., and of Dantzic oak 7490 lbs.

[2] Known in Great Britain as "yellow" pine.

[3] Six tests; other tests varied from 1742 to 4704.

[4] Two tests; a test of Riga white gave 1960 lbs. per sq. in., but this was evidently a defective specimen.

be better appreciated when the number of tests, and the range of strength of the different kinds of wood, are taken consideration. These particulars for thirty-two American woods are given in Table IX, which is copied from Circular No. 15, U.S. Department of Agriculture, Division of Forestry. The difference in the strength of test-pieces of the same species of wood is shown to be very great. The danger of adopting the *average* results of the tests, as sufficiently accurate for practical purposes, is clear from a perusal of the last two columns; in some cases less than one-third of the specimens gave results within 10 per cent of the average strength, and only in a very few cases did all the tests come within 25 per cent of the average. It must be understood that the figures in this table are not the actual crushing strengths of the test-specimens, but represent the strengths corrected to certain percentages of moisture.

TABLE IX.—RESULTS OF TESTS IN COMPRESSION ENDWISE OF THIRTY-TWO AMERICAN WOODS, CORRECTED FOR MOISTURE

[POUNDS PER SQUARE INCH]

No.	Species.	Number of Tests.	Highest Single Test.	Lowest Single Test.	Average Highest 10 per cent of Tests.	Average Lowest 10 per cent of Tests.	Average of all Tests.	Proportion of Tests within 10 per cent of Average.	Proportion of Tests within 25 per cent of Average.
	Reduced to 15 per cent moisture.							Per cent.	Per cent.
1.	Long-leaf Pine (*Pinus palustris*)	1230	11,900	3400	8,600	5700	6,900	53	90
2.	Cuban Pine (*Pinus heterophylla*)	410	10,600	2800	9,500	6500	7,900	61	93
3.	Short-leaf Pine (*Pinus echinata*)	330	8,500	4500	7,600	4800	5,900	47	90
4.	Loblolly Pine (*Pinus tæda*)	660	11,200	3900	8,700	5400	6,500	49	84
	Reduced to 12 per cent moisture.								
5.	White Pine (*Pinus strobus*)	130	8,500	3200	6,800	4000	5,400	49	93
6.	Red Pine (*Pinus resinosa*)...	100	8,200	4300	8,100	4900	6,700	54	96
7.	Spruce Pine (*Pinus glabra*)	170	10,000	4400	8,800	5600	7,300	66	95
8.	Bald Cypress (*Taxodium distichum*) ...	655	9,900	2900	8,500	4200	6,000	31	74
9.	White Cedar (*Chamæcyparis thyoides*) ...	87	6,200	3200	6,000	4400	5,200	79	99
10.	Douglas Spruce (*Pseudotsuga taxifolia*) ...	41	8,900	4100	8,100	4200	5,700	28	65
11.	White Oak (*Quercus alba*)...	218	12,500	5100	11,300	6300	8,500	40	81
12.	Overcup Oak (*Quercus lyrata*)	216	9,100	3700	8,600	6000	7,300	70	95
13.	Post Oak (*Quercus minor*)...	49	8,200	5900	8,100	6000	7,100	58	100
14.	Cow Oak (*Quercus michauxii*)	256	11,500	4600	9,800	5600	7,400	51	89
15.	Red Oak (*Quercus rubra*)	57	9,700	5400	9,200	5500	7,200	36	94
16.	Texan Oak (*Quercus texana*)	117	11,300	5800	9,800	6900	8,100	62	98
17.	Yellow Oak (*Quercus velutina*)	40	8,600	5500	8,300	5800	7,300	58	100
18.	Water Oak (*Quercus nigra*)	31	9,200	6200	9,000	6300	7,800	75	100
19.	Willow Oak (*Quercus phellos*)	153	11,000	4200	8,700	5500	7,200	51	88
20.	Spanish Oak (*Quercus digitata*)	251	10,600	3700	9,500	5100	7,700	61	94
21.	Shagbark Hickory (*Hicoria ovata*) ...	137	13,700	5800	10,900	7500	9,500	79	97
22.	Mockernut Hickory (*Hicoria alba*) ...	75	12,200	6200	11,600	8000	10,100	65	99
23.	Water Hickory (*Hicoria aquatica*) ...	14	10,000	6700	9,600	7000	8,400	71	100
24.	Bitternut Hickory (*Hicoria minima*) ...	25	11,500	7300	11,200	7800	9,600	60	100
25.	Nutmeg Hickory (*Hicoria myristicæformis*)	72	12,300	6400	11,000	7100	8,800	79	97
26.	Pecan Hickory (*Hicoria pecan*)	37	10,500	5800	10,400	7300	9,100	51	95
27.	Pignut Hickory (*Hicoria glabra*)	30	13,000	8700	12,700	8900	10,900	72	100
28.	White Elm (*Ulmus americana*)	18	8,800	4900	8,800	5000	6,500	28	88
29.	Cedar Elm (*Ulmus crassifolia*)	44	10,600	6200	10,100	6500	8,000	66	95
30.	White Ash (*Fraxinus americana*)	87	9,600	5000	8,700	5700	7,200	48	96
31.	Green Ash (*Fraxinus lanceolata*)	10	9,800	6600	9,800	6600	8,000	29	100
32.	Sweet Gum (*Liquidambar styraciflua*) ...	118	8,900	4600	8,500	5600	7,100	60	97

From a consideration of the results recorded in Tables VIII and IX, and of many others, I have come to the conclusion that the highest coefficients of compressive rupture which can safely be adopted in practice, are those adopted in calculating the curves of strength plotted in Plate XXVI. They are intended to apply to timber of reasonably good quality, but not thoroughly seasoned. Green timber, and timber containing serious shakes and numerous knots, must either not be used at all, or must be employed with a larger factor of safety than is necessary for good material. The curves of strength have been calculated according to the author's formula. By means of these curves the approximate ultimate strength of rectangular posts, up to 70 diameters high, can be obtained in a moment. Thus, if it is required to find the strength of a red deal post 12 inches by 6 inches and 20 feet high, find first the ratio between the length and the least diameter (*i.e.* $\frac{20 \text{ feet} \times 12}{6} = 40$), look in one of the vertical columns at the sides of the diagram for this ratio, and follow the horizontal line from this till it cuts the curves for red deal; the crushing strength in lbs. per square inch can then be read at the head or foot of the diagram, namely, 1150 to 1470 lbs. per square inch. If the timber is of good quality the higher value may be taken, and the crushing strength of the column will therefore be 12 inches × 6 inches × 1470 lbs. = 105,840 lbs. With a factor of safety of 5 the safe load will be 21,168 lbs.

Some experiments on the strength of posts, made up of two or three scantlings fixed side by side, have been carried out at the Watertown Arsenal, and are summarized in the following table:—

TABLE X.—STRENGTH OF SINGLE AND COMPOUND POSTS

Wood.	Description.	Ratio, $\frac{l}{d}$	No. of Tests.	Ultimate Strength in lbs. per sq. in.
Yellow Pine	Single posts	16	6	4590
" "	" "	32·7	3	3472
" "	Posts with 2 scantlings	16	12	3722
" "	" " 3 "	15·5	6	3122
White Pine	Single posts	16	6	2345
" "	" "	21·4	9	2326
" "	" "	40	3	1672
" "	" "	32·4	3	2431
" "	Posts with 2 scantlings	20	3	1785
" "	" " 2 "	16	15	1972
" "	" " 3 "	15·5	6	1962
" "	" " 4 "	15·5	6	1972

The general deduction from these tests is that, for all practical purposes, the strength of a compound column may be taken to be the sum of the strength of the separate pieces. In other words, if a compound column 12 feet long is made up of (say) two 11-inch × 4-inch pieces, each piece will have a ratio of diameter to length = 1:36, and the strength of a post with this ratio must be calculated to give the ultimate strength per square inch which the compound column may be expected to attain. If the ratio of the compound column itself (*i.e.* in this case, 1:18) is taken as the basis of calculation, too high results will be obtained.

The strength of circular posts will be approximately the same as that of square posts *of the same area.* The side of a square, whose area shall be equal to that of a given circle, is found by multiplying the diameter of the circle by ·882. Thus, if it is required to find the strength of a circular post 12 inches in diameter, find the side of a square of equal area by multiplying 12 by ·882; this gives roughly 10·6 inches, and the strength of a square post of this size (the height being known) can be calculated directly by the formula, or deduced from the curves of strength shown graphically in Plate XXVI.

We have seen that green timber is very much weaker than seasoned, but as it may be assumed that such timber will in the course of time attain the normal strength, it would not call for a much higher factor of safety were it not for the fact that shrinkage and twisting will in all probability occur during the process of drying, and cause faulty bearings at the ends, and consequently eccentric loading. It is therefore important that timber should be at least fairly dry before being used in construction.

Eccentric loading reduces the strength of a post very considerably. Some experiments at the Watertown Arsenal seem to show that a load applied about 2 to $2\frac{3}{4}$ inches off the axis of posts from 8 to 11 inches in diameter reduced the strength about 20 per cent. This emphasizes the necessity of good workmanship and firm foundations.

In carpentry, compression members are usually framed into other pieces of timber, and this is another source of weakness. Professor Lanza mentions the case of a yellow-pine column with a sectional area of 68·8 square inches and a length of 12 feet 6·85 inches, which was tested with one end resting on a thick yellow-pine bolster. The post began to split under a load of 1744 lbs. per square inch, owing to eccentricity of bearing caused by the uneven yielding of the bolster. The split portion ($2\frac{1}{2}$ inches in length) was cut off, and the post tested without a bolster; it did not fail until a load of 5451 lbs. per square inch had been reached. The following table has been prepared from Professor Lanza's records of his tests at the Watertown Arsenal, in order to show at a glance the effect of different kinds of bearing. The posts varied in diameter from 5·85 to 10·56 inches, and in length from 11 feet 6·2 inches to 14 feet.

Table XI.—TESTS OF OLD AND SEASONED WHITE-OAK POSTS

No.	Method of Testing.	Length / Diameter	No. of Tests.	Average Ultimate Strength in lbs. per sq. in.
1.	Ends brought to even bearing ...	20·5	3	5079
2.	Ends not brought to even bearing	17·2	8	4233
3.	Tested with maple cap and oak base	24	4	3469
4.	Iron cap and pintle at one end, and iron base at the other	13·8	1	4838

These tests show clearly the weakening effect of uneven bearings, and of wood bolsters. The same lesson is taught by the experiments at the Massachusetts Institute of Technology, referred to on page 349. The cause of the early failure of posts resting on wood bolsters is to be found in the uneven crushing of the latter, and the consequent uneven bearing at the ends of the posts.

Wood loaded across the grain, which is what occurs in the case of bolsters, begins to fail much sooner than when crushed in the direction of the grain. The U.S.A. Department, whose tests of compression endwise have been given in Table IX, also made numerous experiments on compression across the grain, the results (corrected for moisture as in Table IX) being as follows, in lbs. per square inch, "to an indentation of 3 per cent of the height of the specimen":—white ash, 1900; cedar elm, 2100; white elm, 1200; sweet gum, 1400; pecan hickory, 2800; pignut hickory, 3200; shagbark hickory, 2700; white oak, 2200; red oak, 2300; white cedar, 700; bald cypress, 800; Cuban, loblolly, long-leaf, and red pines, 1000; short-leaf pine, 900; spruce pine, 1200; white pine, 700; Douglas spruce, 800.

The importance of firm and accurate bearings for wood posts will be appreciated by a consideration of the facts just recorded. A great increase of strength will be secured if the ends of the posts are accurately fitted with metal caps or sockets, provided with flanges so as to cover a wider area of the beams or bolsters on which the posts are resting. An architect or engineer spreads out the base of an iron or steel column resting on brickwork in order to distribute the weight over a wider area, and a corresponding arrangement will prove advantageous in timber structures.

It is unnecessary to consider in detail the strength of columns with one or both ends rounded (or, as it is commonly expressed, "hinged"), as cases of this kind do not occur in practice, unless the workmanship is of the most careless kind. It may, however, be mentioned that rounded ends reduce very considerably the strength of columns. Kirkaldy, in his *Strength and Properties of Materials*, gives tests of deodar pine (or Himalayan cedar) with flat and rounded ends; twelve specimens with flat ends had an average strength of 4265 lbs. per square inch, while the average of six with rounded ends was only 1970 lbs. The former specimens, however, were only 5 diameters high, while the latter were 20, but this only accounts for a small portion of the difference. Calculated according to the new formula, the longer columns ought to have had a strength of 3392 lbs. per square inch. If this is correct the strength was reduced more than 40 per cent by rounding the ends.

CHAPTER IV

TRANSVERSE STRENGTH AND ELASTICITY

Tests of the transverse strength of timber are usually made on beams supported at the ends and loaded at the centre, and the formula for rectangular beams, supported and loaded in this manner, is—

$$\text{w} = \frac{2 f b d^2}{3 l} \quad \text{(3, page 317)}$$

Where w = breaking weight in lbs.,
f = coefficient of bending strength, or modulus of rupture, in lbs. per square inch,
b = breadth of the beam in inches,
d = depth „ „
l = length „ „ (between the supports).

To find the value of f from this formula, it is necessary to test a number of beams by supporting them at the ends and loading them in the centre till they break. The breaking weight of these test-specimens being found, the value of f can be calculated by transposing the formula, thus:—

$$f = \frac{3 \text{w} l}{2 b d^2}$$

As an example, we may mention three tests of 9 inches by 3 inches Archangel deals, 10 feet long between the supports, which broke with an average central load of 8442 lbs. These give us the following data:—

w = 8442 lbs., b = 3 inches, d = 9 inches, and l = 120 inches.

$$\text{Therefore } f = \frac{3 \times 8442 \times 120}{2 \times 3 \times 9^2} = 6253 \text{ lbs. per square inch.}$$

The tests just recorded were carried out by Mr. Kirkaldy, but in Mr. W. G. Kirkaldy's *Strength and Properties of Materials* the value of the coefficient is worked out from the formula $\text{s} = \frac{\text{w} l}{4 b d^2}$, which makes s = 1042, or one-sixth of the value of f. In this case s must be regarded as a "constant", and not as the true coefficient of rupture.

In comparing the results obtained by different experimentalists, it is therefore essential to know by what formula the value of the constant or coefficient of rupture has been calculated, otherwise serious errors may be made. The accepted theory of the strength of beams, which includes the theory of bending moments and of moments of resistance, necessitates the equations given in Chapter III of the first part of this section.

These formulas are not, however, accurate when applied to breaking tests, as they assume that the neutral plane remains constantly at the centre of the depth of the rectangular beam. This assumption is incorrect, as timber is much stronger in tension than in compression, and before rupture the neutral plane is lowered. In the case of green timber (which has relatively a very small compressive strength) the plane may be lowered to three-fourths or more of the depth of the beam. It follows, therefore, that the value of f ascertained in this manner requires rectification, but to what extent it is impossible to state. The difficulty is usually met by adopting a value considerably below the average obtained from the tests.

Early experimentalists confined themselves to tests of small pieces of wood, many of them only 1 inch square, and as these were generally straight-grained, well-seasoned, and free from knots and other defects, they often gave results higher than the average which can be obtained from large scantlings, although the difference is not so marked as in tests of compressive strength. It is wise, however, to ignore the tests of small specimens, or to correct the results in the light of experiments carried out on full-sized beams. Figures are not yet available for every variety of timber, but a great number of tests are collected in Table XII. The most trustworthy tests of American woods are those carried out by the U.S.A. Division of Forestry, and recorded in column 5. Professor Lanza's tests (L, column 3) are also of great value, as the figures given are the averages of a very large number of specimens of ordinary scantling.

Table XII.

TRANSVERSE STRENGTH OF TIMBER,—COEFFICIENTS OF RUPTURE, Columns 1–7 $\left(f=\frac{3\,wl}{2\,bd^2}\right)$, AND MODULUS OF ELASTICITY, Column 8 $\left(E=\frac{W}{D}\times\frac{l^3}{4bd^3}\right)$

For the botanical names of the timbers, see Table VIII, pages 349 and 350.

1. Laslett's tests of small specimens, 2″ × 2″ and 6 ft. span.
2. K.=Kirkaldy; C.=Edwin Clark and C. Graham Smith; W.=Professor Warren, specimens about 6 in. by 3½ in. by 4 ft. 6 in.
3. L.=Lanza; B.=Bauschinger; S.=Swedish timber (Wijkander), 3·94 ins. square; U.=Prof. Unwin, small specimens.
4. R.=H. P. Robertson (Western Australian timber, mean between tests by J. A. M'Donald and Admiralty of W. Australia); W.=Woolwich experiments on scantlings, 6″ × 3¼″ × 14′ 3½″ (Seddon), and (for Yellow or Red Deal) 9″ × 3″, 9″ × 6″, and 12″ × 9″.
5. Tests of "green" timber, U.S.A. Division of Forestry, corrected to 15 per cent moisture for Cuban, Loblolly, Long-leaf and Short-leaf Pines, and 12 per cent for other woods. **6.** Rankine's recommendations, *mean.*
7. Recommendations of a committee of the American Association of Railway Superintendents of Bridges and Buildings.
8. Modulus of Elasticity. Unless otherwise stated, the figures are those given in Circular No. 15, U.S. Dep. Agric., Div. Forestry. B.=Bauschinger; K.=Kirkaldy; L.=Lanza; R.=H. P. Robertson, S.=Swedish timber, according to Wijkander; W.=Woolwich Arsenal.

	1.	2.	3.	4.	5.	6.	7.	8.[1]
Hardwoods.								
Acacia (Blackwood)	...	11,091, W.	12,220, U.	...	...	...	...	{ 1,618,000, W. { 2,383,000, U.
Alder	...	...	9,143, S.	...	...	...	...	1,479,000, S.
Ash, American white ...	8,600	...	12,122, L.	...	10,800	...	...	1,640,000
Common	11,600	...	10,707, S.	...	...	13,000	...	...
Baywood (see Mahogany).								
Beech	...	...	13,210, S.	...	...	10,500	...	1,720,000, S.
Birch, Silver	...	...	10,835, S.	...	...	11,700	...	1,621,000, S.
Yellow	...	...	7,930,[2] L.	...	...	...	...	1,797,000, L.
Chestnut, American	...	...	...	...	...	...	5,000	...
"	...	...	...	...	...	10,660	...	...
Ebony	...	...	...	...	...	27,000	...	...
Elm, Cedar	...	...	...	...	13,500	...	...	1,700,000
Common	5,300	...	8,731, S.	...	...	7,850	...	1,095,000, S.
White	12,400	...	...	...	10,300	...	...	1,540,000
Eucalyptus, Blue Gum[3] ...	9,600	13,255, W.	8,006, U.	...	...	18,000	...	{ 1,747,000, W. { 832,000, U.
Ironbark	19,000	18,000, W.	15,790, U.	...	...	...	...	2,600,000, W.
Jarrah	9,200	12,040, W.	...	8,900, R.	...	...	...	{ 2,080,000, R. { 1,390,000, W.
Karri	11,600	11,208, W.	...	8,000, R.	...	...	...	2,002,000, W.
Tallow-wood, &c.	...	14,370, W.	12,289, U.	...	...	...	...	2,120,000, W.
Tuart	...	...	...	9,300, R.	...	...	...	2,300,000, R.

[1] For practical use, the values of E ought to be taken as one-half of those recorded below; see pp. 358–9. [2] Mean of two tests.
[3] Warren's tests of Blue Gum from Victoria, Unwin's from Queensland.

	1.	2.	3.	4.	5.	6.	7.	8.
HARDWOODS—*Cont.*								
Greenheart	17,900	...	19,836, U.	...	...	...	...	2,735,000, U.
Gum, Sweet	...	...	...	...	9,500	...	...	1,700,000
Hemlock	...	...	3,825, L.	...	...	...	...	922,000, L.
Hickory, Pecan	...	...	...	...	15,300	...	...	2,530,000
Pignut	...	...	...	...	18,700	...	...	2,730,000
Shagbark	...	...	...	...	16,000	...	...	2,390,000
Hornbeam	...	...	12,584, S.	...	...	...	...	1,749,000, S.
Lignum-vitæ	...	...	...	...	...	12,000	...	...
Lime	...	...	9,143, S.	...	...	...	...	867,000, S.
Mahogany, African	14,900	...	...	...	...	14,980	...	...
Baywood	10,800	...	...	...	...	} 9,550	...	...
Spanish	11,500	...	...	...	...			
Maple	...	...	7,146,[1] L.	...	...	...	...	1,517,000, L.
„ Sycamore	...	...	...	...	...	9,600	...	...
Morawood	...	...	18,251, U.	...	...	...	...	2,837,000, U.
Oak, American white ...	10,800	7,662, K.	5,863,[2] L.	...	13,100	...	6,000	{ 1,131,000, L. 2,090,000
„ red	...	...	...	...	11,400	10,600	...	1,970,000
European	9,300	9,243, K.	10,670, S.	...	...	10,760	...	1,436,000, S.
Poplar, Aspen	...	...	9,157, S.	...	...	...	...	1,251,000, S.
Sabicu	17,800	...	...	...	...	...	...	...
Teak	12,300	...	...	...	...	15,500	...	...
Willow	...	...	...	...	...	6,600	...	...
SOFTWOODS.								
Cedar, Lebanon	...	...	...	...	...	7,400	...	...
White	...	...	...	...	6,300	...	...	910,000
Variety not stated ...	7,500	...	...	...	...	...	5,000	...
Cypress, Bald	...	...	...	...	7,900	...	...	1,290,000
Not stated	...	...	...	...	...	...	5,000	...
Hemlock	...	...	3,825,[3] L.	...	...	...	3,500	...
Larch	8,400	...	{ 9,555, B. 6,370, S. }	...	...	7,500	...	952,000, S.
Pine, Cuban	...	...	...	...	11,900	...	...	2,300,000
Kauri	11,000	...	4,845, U.	7,330, W.	...	11,000	...	1,052,000, U.
Loblolly	...	...	...	...	10,100	...	...	1,950,000
Long-leaf	...	...	...	...	10,900	...	7,000	1,890,000
Pitch, &c.	14,100	{ 6,984, C. 7,626,[4] K. }	...	...	...	...	...	...
Red, &c.	8,800	4,707, C.	...	8,849, W.	4,900	8,320	5,000	1,620,000
Short-leaf	...	...	...	...	9,200	...	6,000	1,600,000
Spruce	...	...	...	...	10,000	...	...	1,640,000
White	...	...	4,451,[5] L.	...	7,900	...	4,000	{ 1,390,000 1,122,000, L.
„ (Canadian) ...	6,800	4,491, C.	...	...	...	...	...	...
Yellow, &c.	...	...	7,442,[6] L.	...	...	...	...	1,783,000, L.
„ or Red Deal, &c.	...	...	7,696,[7] B.	4,218, W.	...	...	...	1,174,000, W.
Beams ...	...	{ 5,361, C. 4,642, K. }	...	3,468, W.	...	...	...	1,016,000, W.
Deals and battens, Russian	8,100	6,846, K.	...	...	...	...	...	1,915,000, K.
Do. Swedish ...	...	6,341, K.	9,442, S.	...	...	...	...	{ 1,505,000, K. 1,592,000, S.
Sequoia	...	...	...	...	...	...	4,500	...
Spruce, American	9,000	...	4,521,[8] L.	...	...	...	4,000	1,310,000, L.
California	...	...	...	...	...	...	4,500	...
Douglas	...	...	...	...	7,900	...	...	1,680,000
Whitewood	...	...	{ 7,249,[9] B. 7,588, S.	...	...	...	...	{ 1,407,000, B. 1,507,000, S.

[1] Six tests, chiefly about 4 in. × 12 in.; highest 11,080, lowest 4260.

[2] Average of 36 tests from about 3″ × 11″ to 6″ × 12″; highest 8850, lowest 3240.

[3] Average of 17 tests from about 3″ × 9″ to 4″ × 12″; highest 6535, lowest 1531.

[4] Mean of 10 beams from about 11 to 13 ins. square.

[5] Average of 37 tests from about 3″ × 9″ to 6″ × 12″; highest 7251, lowest 2456. Seven tests of kiln-dried Western White Pine gave an average of 5482.

[6] Average of 99 tests from about 3 ins. × 12 ins. to 6 ins. × 16 ins.; highest 11,360, lowest 3828.

[7] About 15% moisture, therefore well seasoned.

[8] Average of 160 tests from about 2 ins. × 9 ins. to 8 ins. × 12 ins.; highest 9036, lowest 1794.

[9] About 17% moisture.

The ultimate strength of a beam is not, however, all that needs to be considered. A beam may be strong enough to bear safely the load which is placed upon it, but may bend or "deflect" so much as to interfere with the utility and sightliness of the structure. Some kinds of timber have a high ultimate strength, but a comparatively small resistance to deflection; they are strong, but not stiff. Thus, in the preceding table, the coefficient of rupture of red or Norway pine, according to the tests of the U.S.A. Division of Forestry (column 5), is 4900, and that of short-leaf pine very nearly twice as much, namely 9200, but according to the same authority the modulus of elasticity of the two materials is practically the same, that of red pine being indeed slightly the higher. In other words, red pine is as stiff as short-leaf pine, although its ultimate strength is only about one-half. In designing timber beams it is therefore necessary to consider, not only the breaking weight, but the amount of deflection which will be produced by the given load.

We need here consider only the cases of cantilevers and of beams supported at both ends. The formulas[1] are as follows:—

1. *Beam fixed at one end and loaded at the other.*

$$v = \frac{4\,\mathrm{w}\,l^3}{\mathrm{E}\,b\,d^3}, \text{ and } \mathrm{E} = \frac{4\,\mathrm{w}\,l^3}{v\,b\,d^3} \qquad (24)$$

Where v = greatest deflection in inches,
E = modulus of transverse elasticity of the material in lbs. per square inch,
W = the total load in lbs.,
l = length of span in inches,
b = breadth of beam in inches,
d = depth of beam in inches.

2. *Beam fixed at one end and loaded uniformly.*

$$v = \frac{3\,\mathrm{w}\,l^3}{2\,\mathrm{E}\,b\,d^3}, \text{ and } \mathrm{E} = \frac{3\,\mathrm{w}\,l^3}{2\,v\,b\,d^3}. \qquad (25)$$

3. *Beam supported at both ends and loaded in the centre.*

$$v = \frac{\mathrm{w}\,l^3}{4\,\mathrm{E}\,b\,d^3}, \text{ and } \mathrm{E} = \frac{\mathrm{w}\,l^3}{4\,v\,b\,d^3}. \qquad (26)$$

4. *Beam supported at both ends and loaded uniformly.*

$$v = \frac{5\,\mathrm{w}\,l^3}{32\,\mathrm{E}\,b\,d^3}, \text{ and } \mathrm{E} = \frac{5\,\mathrm{w}\,l^3}{32\,v\,b\,d^3} \qquad (27)$$

From the second formula in each case the modulus of transverse elasticity of any material can be calculated by observing the amount of deflection under a certain load. This load, however, must not be the breaking load or anywhere near it, as when the breaking load is approached the deflection increases out of all proportion, and a very low modulus of elasticity will be obtained if it is calculated from the maximum deflection recorded just before rupture. This will be better understood by reference to an actual test. A scantling of long-leaf pine, supported at both ends and loaded at the centre, with a length of 140 inches between the supports, a breadth of 4·02 inches, and a depth of 8·04 inches, showed the following deflections under increasing loads:—0·17 inch at 1000 lbs., 0·34 at 2000, 0·50 at 3000, 0·66 at 4000, 0·82 at 5000, 0·96 at 6000, 1·13 at 7000, 1·27 at 8000, 1·46 at 9000, 1·65 at 10,000, 1·93 at 11,000, 2·27 at 12,000, 2·85 at 13,000, 3·85 at 13,500. The beam broke under this load.

It will be observed that the last 500 lbs. increased the deflection no less than 1 inch, or as much as the deflection caused by the first 6000 lbs. of the load. If the results are plotted graphically, as in fig. 577, we can see at a glance that, up to 9000 lbs., the deflection

[1] Professor Norton has suggested a more elaborate formula, but those here given are generally accepted as sufficiently accurate for all practical purposes.

increased directly as the load, but that, beyond this point, the deflection increased in a much greater proportion. The correct modulus of elasticity may be calculated from any of the deflections recorded up to the load of 9000 lbs. In this case, therefore, the limit of transverse elasticity is about two-thirds of the breaking weight, 9000 lbs. being two-thirds of 13,500 lbs., the weight under which rupture occurred.

For a beam supported and loaded in this manner, $E = \frac{wl^3}{4vbd^3}$; to find the value of E we must base our calculation on the deflection under any load not exceeding 9000 lbs. Suppose we take the deflection of 1·27 inch under a load of 8000 lbs.; then

$$E = \frac{8000 \times 140^3}{4 \times 1{\cdot}27 \times 4{\cdot}02 \times 8{\cdot}04^3} = 2{,}070{,}000 \text{ lbs. per square inch.}$$[1]

A practically similar result will be obtained by working from any other load not exceeding 9000 lbs.; but if higher loads are considered, the modulus of elasticity will be much lower. Thus, if the calculation is based on the deflection of 3·85 inches under a load of 13,500 lbs., E = 1,151,000 lbs. per square inch, or little more than half the true modulus. The most correct method of determining the modulus is to take the average deflection per 1000 lbs. load for all loads up to the elastic limit of the material; in this case the average is ·162 inch per 1000 lbs., and it will be found that E = 2,026,000 inch-lbs.

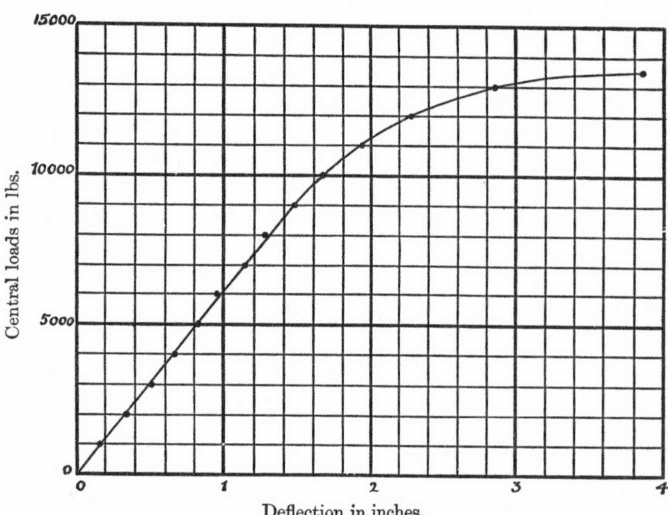

Fig. 577.—The Deflection of a Long-leaf Pine Beam under increasing Loads

Different methods of calculation have been adopted by different experimentalists, and there is consequently a greater discrepancy between the calculated moduli of transverse elasticity than the tests really warrant. Many kinds of wood have not yet been satisfactorily tested, but the values of E for some of the most important woods are given in column 8, Table XII.

These values, however, require modification. Some of them are averages of a few tests, others of a considerable number, but all averages are dangerous when used as the basis of calculation. The average value of E obtained by Professor Lanza for American spruce is given in the table as 1,310,000, but many of the individual tests gave very much lower results, five (out of 143 tests) being below 900,000, and the lowest being only 849,500, or less than two-thirds of the average. The lowest test of yellow pine gave a value of 1,058,200, but the average value was 1,783,000. White oak had an average of 1,131,100, and a lowest value of 672,724; white pine an average of 1,122,000, and a lowest value of 727,200; hemlock an average of 922,250, and a lowest value of 412,670.[2]

It is necessary also to point out that the deflection of timber increases considerably under a long-continued load. We have all known instances of roof-purlins which have borne their loads without appreciable deflection for months, but have eventually sagged to an unsightly extent, although the loads have not been increased. When we remember that tests of timber are usually made quickly, we are disposed to ask whether the moduli of elasticity thus obtained are not too high for practical use. Some experiments by Professor Lanza on spruce and yellow-pine beams show that the answer to this question must be in

[1] Approximately; in stating the values of moduli of elasticity it is customary to ignore the hundreds, tens, and units.

[2] This was exceptionally low, and the beam must have been very defective; the next lowest test was 688,960.

the affirmative. These beams were about 4 inches broad, 12 inches deep, and 18 or 20 feet long between the supports, and were seasoned for about six months, some under a central load and others without load. The deflection of the loaded beams was carefully recorded. After seasoning, all the beams were kept for a few days without load until they were finally tested. The modulus of elasticity calculated from the original deflection of the loaded beams was, after allowing for the shrinkage during seasoning, practically the same as that obtained from the other beams, and the latter need not therefore be further considered.

The results of the time-tests of the loaded beams are grouped together in the following table, which is a short summary of four tables in Professor Lanza's *Applied Mechanics*:—

TABLE XIII.—EFFECT OF LONG-CONTINUED LOADS ON THE TRANSVERSE ELASTICITY OF TIMBER

Description of Wood.	No. of Tests.	E (immediate).	E (final).	Reduction per cent.
		lbs. per sq. in.	lbs. per sq. in.	
1. Spruce from Maine	6	1,300,000	963,500	26[1]
2. Do.	6	1,376,500	614,000	55
3. Do.	6	1,307,500	614,000	53
4. Yellow Pine from Georgia ...	6	1,729,000	867,000	50

From these tests it would appear that in practice the modulus of elasticity ought to be assumed to be *not more than one-half* that obtained by quick laboratory tests. It is possible that the deflection would not have increased to such an extent if the beams had been well seasoned before the first application of the load. This, however, is by no means certain. The beams in series 1 and 3 were all green with the exception of one in series 1, referred to in the note to the table. Those in series 2 were cut in the spring of 1886, and first loaded in the middle of September. The yellow-pine beams were cut in the season of 1886, and did not reach the testing laboratory till September, 1897. It would appear, therefore, that until further tests have been made, the values of E, for use in practice, ought to be *one-half* only of the values given in column 8, Table XII.

The illustration on page 358 shows that even under the smallest load timber deflects, and a certain amount of deflection must therefore always be allowed for. Opinions differ somewhat as to the maximum permissible deflection. In America a deflection of $\frac{1}{300}$ to $\frac{1}{400}$ of the span is allowed for supported beams. Much, however, depends upon the nature of the structure. The following figures will give satisfactory results:—

1. *For supported beams*—

	Permissible Deflection.
a. Carrying ceilings	$\frac{1}{480}$ of the span.
b. Without ceilings	$\frac{1}{360}$ "
c. Temporary structures	$\frac{1}{250}$ "

2. *For cantilevers*—

	Permissible Deflection.
a. Carrying ceilings	$\frac{1}{960}$ of the span.
b. Without ceilings	$\frac{1}{720}$ "
c. Temporary structures	$\frac{1}{500}$ "

If we revert again to the actual test recorded graphically in fig. 577, we find that the beam had a clear span of 140 inches. If we consider this as a floor-joist supporting a ceiling, the maximum permissible deflection will be $\frac{1}{480} \times 140$ inches = ·29 inch, and it will be seen from the diagram that this deflection was reached under a load of about 3600 lbs., or rather more than one-fourth of the breaking weight. But if, in view of Professor Lanza's time-tests, we are to assume in practice only one-half the stiffness recorded by quick laboratory tests, we must conclude that a deflection of $\frac{1}{480}$ of the span will ultimately ensue from a load of only 1800 lbs., or between one-seventh and one-eighth of the breaking weight. This ratio will, of course, vary for different kinds of timber, but the example will suffice to

[1] In this series, five of the beams (loaded when green) showed an average reduction of 50 per cent, but the sixth beam, which had been seasoned six months before being loaded, gave a greater value of E at the end of the test than at the beginning.

show that the stiffness of a beam is as important a basis of calculation in many cases as the strength.

We have seen that the strength of a beam increases as the square of the depth, and that the stiffness increases as the cube of the depth; thus, a beam 8 inches deep is four times as *strong* as one 4 inches deep, but eight times as *stiff*. It would appear from this that the stiffest and most economical beam would be that with the greatest proportionate depth. This, however, is only true within certain limits, as a beam which is too deep will fail by longitudinal shearing. The shearing force for some methods of loading is easily calculated by formulas; for more complicated loads, reference must be had to the graphic solutions given in Chapter III, Part I, of this Section. Let F = shearing force in lbs. at any vertical section of the beam, w = load per unit of length l, W = single load, and wl = W; then

1. *For beams fixed at one end and free at the other.*

a. Loaded at the free end, F = W. .... (28)
b. Loaded uniformly, F increases from 0 at the free end to wl (= W) at the fixed end.......... (29)

2. *For beams supported at both ends.*

a. Loaded at the centre, $F = \frac{W}{2}$, except at the centre of the span, where F = 0.......... (30)

b. Loaded uniformly, F increases from 0 at the centre of the span to $\frac{wl}{2}\left(=\frac{W}{2}\right)$ at each support...... (31)

The greatest shearing force is therefore the same whether the beam is loaded centrally or uniformly, and the same formula can be used for calculating the greatest intensity of the shearing force, which will be at the neutral axis of the section. Let S = the greatest intensity of the shearing force in lbs. per square inch, W = the total load in lbs., b = the breadth of the beam, and d = the depth; then

1. *For beams fixed at one end and free at the other.*

$$S = \frac{3W}{2bd} \qquad (32)$$

2. *For beams supported at both ends.*

$$S = \frac{3W}{4bd} \qquad (33)$$

Example.—A spruce beam 3 inches wide and $11\frac{7}{8}$ inches deep, supported at both ends, was tested by Professor Lanza, and failed by shearing under a load of 8927 lbs. Find the greatest intensity of longitudinal shearing force. By formula (33), we have

$$S = \frac{3 \times 8927}{4 \times 3 \times 11\frac{7}{8}} = 187 \text{ lbs. per square inch,}$$

which is the answer required.

Out of 159 spruce beams tested by Professor Lanza, 24 failed by shearing, or by shearing in conjunction with crushing and tensile rupture, the average greatest intensity of shear at the point of failure being 196 lbs. per square inch. Out of 99 tests of yellow-pine beams 24 failed in this manner, the average greatest intensity of shear being 246 lbs. per square inch. Two white-oak beams, out of 36 tested, failed in the same way, the shear being 152 and 379 lbs. per square inch respectively, or an average of 266. Three white-pine beams, which failed out of 37 tested, had an average shearing strength of 151 lbs. per square inch. The individual tests varied very widely, from 117 to 258 for spruce, 151 to 445 for yellow pine, 152 to 379 for oak, and 119 to 180 for white pine. It is clear, therefore, that the shearing strengths of these woods ought in practice to be assumed to be somewhat lower than the averages just given, even though these are much lower than the shearing strengths recorded in Table VI, page 340. The reason is, that in the specimens tested for shearing, rupture was made to occur along a plane of sound wood, whereas in the beams shearing took place along the weakest plane, which might be seriously damaged by shakes and other defects.

For the purpose of comparing the transverse strength, deflection, and shearing resistance of beams of different proportions, we will consider in detail three beams containing exactly the same amount of wood, namely A, 6 inches by 6 inches; B, 3 inches by 12 inches; and (as an extreme) C, 2 inches by 18 inches; all the beams to be 15 feet or 180 inches between the supports, to be centrally loaded, and to be of Cuban yellow pine with a coefficient of rupture of 8000 and a modulus of elasticity of (say) 2,000,000.

To find the breaking weight we must use the formula, $W = \frac{2fbd^2}{3l}$. For beam A we shall have

$$W = \frac{2 \times 8000 \times 6 \times 36}{3 \times 180} = 6400 \text{ lbs.}$$

To find the load which will produce a deflection not exceeding (say) $\frac{1}{360}$ of the span —*i.e.* not exceeding $\frac{1}{2}$ inch—we must adopt the formula, $v = \frac{Wl^3}{4Ebd^3}$, transposed thus, $W = \frac{4vEbd^3}{l^3}$. For beam A we shall therefore have

$$W = \frac{4 \times \frac{1}{2} \times 2{,}000{,}000 \times 6 \times 216}{5{,}832{,}000} = 888 \text{ lbs.}$$

To find the greatest intensity of shearing force produced by the load causing the maximum permissible deflection, namely 888 lbs. in the case of beam A, we apply the formula, $S = \frac{3W}{4bd}$.

$$S = \frac{3 \times 888}{4 \times 6 \times 6} = 18\tfrac{1}{2} \text{ lbs.}$$

Proceeding in a similar manner for beams B and C, we get the breaking weights, the loads producing the greatest permissible deflection, and the greatest shearing force produced by the deflection loads, as recorded in Table XIV.

TABLE XIV.—COMPARISON OF THREE CUBAN YELLOW-PINE BEAMS CONTAINING THE SAME AMOUNT OF TIMBER

	Clear Span in ins.	Breadth in inches.	Depth in inches.	Breaking Weight in lbs.	Ratios.	Load producing $\frac{1}{2}$ in. Deflection, in lbs.	Ratios.	Maximum Shearing Force under the Breaking Weight, in lbs.	Ratios.	Maximum Shearing Force at the Deflection Limit, in lbs.	Ratios.
Beam A ...	180	6	6	6,400	1	888	1	133	1	18·5	1
„ B ...	180	3	12	12,800	2	3555	4	266	2	74	4
„ C ...	180	2	18	19,200	3	8000	9	400	3	166·6	9

Beam A is manifestly weak and flexible; a deflection equal to $\frac{1}{360}$ of the span is produced by a load of less than $\frac{1}{7}$ of the breaking weight, and this load is the greatest allowable load for the beam. The shearing stress is so small that it need not be taken into consideration.

Beam B is twice as strong as A, and four times as stiff. In this case it would be unwise to load the beam up to the deflection limit, as this load is only about $\frac{2}{7}$ of the breaking weight, giving therefore a factor of safety of only $3\frac{1}{2}$. The shearing stress at the deflection limit is between $\frac{1}{3}$ and $\frac{1}{4}$ of the average stress at which Professor Lanza's yellow-pine beams failed, and therefore allows a sufficient margin of safety.

Beam C is three times as strong as A, and nine times as stiff. This beam would be in danger of failing by shearing, even at the deflection limit, and would almost certainly fail by shearing before the theoretical breaking weight was reached; the objection to loading it up to the deflection limit is greater than in the case of beam B for the reason just given, and also because the factor of safety would be less than $2\frac{1}{2}$. Such a deep and narrow beam would also be likely to fail by buckling before the theoretical breaking load is reached, as in practice it is almost impossible to avoid some eccentricity of loading.

From these examples it will be evident that, in ordinary cases, it is not necessary to take the shearing stresses into consideration, as in beams of common scantling, where the depth is not more than three or four times the breadth, the shearing stresses will be reduced to safe limits by making the working load not more than one-fifth of the breaking load—in other words, by calculating the breaking load and dividing it by a factor of safety (not less than 5) to ascertain the safe working load. In beams of greater relative depth, however, the shearing force becomes so great as to call for proper calculation. It is also clear that in beams of small relative depth the working load must be calculated from the permissible deflection rather than from the breaking weight, as the latter, divided by an ordinary factor of safety, would give a working load which might deflect the beam beyond the limit allowed.

The *strongest* rectangular beam, which can be cut out of a round log, has its cross-section of the following proportion:—

depth : *breadth* : : $\sqrt{2}$: 1, or approximately as 7 is to 5;

while the *stiffest* rectangular beam, which can be cut out of a round log, is proportioned thus:—

depth : *breadth* : : $\sqrt{3}$: 1, or approximately as 19 is to 11.[1]

These ratios are sometimes misunderstood, being taken to mean that a beam with depth and breadth in the proportion of 7 and 5 is stronger than any other rectangular beam containing the same amount of timber, and that a beam proportioned as 7 is to 4 is stiffer than any other rectangular beam containing the same quantity of material. This, of course, is far from correct. The proportions are merely applicable to cases where it is required to cut the strongest or stiffest single beam from a given log. If we have a log $17\frac{1}{4}$ inches in the least diameter, the strongest beam which can be cut from it will be as shown at A B C D, fig. 578, and will measure 14 inches by 10 inches, while the stiffest beam will be about 14·9 inches by 8·6 inches, as shown at E F G H. The largest square beam which can be cut out of the log will measure 12·19 inches on the side, and will contain more timber than either of the other beams, but will be neither as strong nor as stiff.

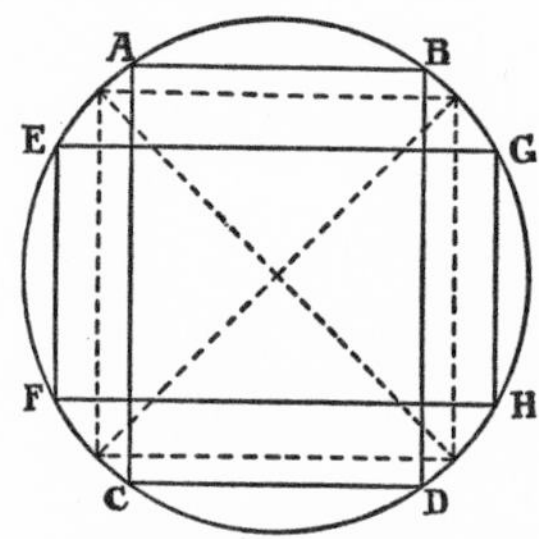

Fig. 578.—Strongest and Stiffest Beams which can be cut from Cylindrical Log

It is probable that for the same kind of timber there is a relation between the coefficient of rupture (strength) and the modulus of elasticity (stiffness), but timber is such an uncertain material that the relation will probably never be closely defined. Bauschinger, from his tests of pine, larch, and fir, came to the conclusion that, for these materials, cross-breaking strength (in lbs. per square inch) = ·0045 modulus of elasticity (in lbs. per square inch) + 450, or, using the letters adopted in preceding formulas—

$$f = \cdot 0045\,E + 450 \qquad (34)$$
$$\text{and } E = 222\cdot\dot{2}(f - 450) \qquad (35)$$

Professor Lanza's tests of spruce and other timbers show that some relationship between strength and stiffness does exist, although this relationship does not accord with Bauschinger's formula. If we take the average of a certain number of the highest tests of *strength* (namely, the values of f) and the average of the same number of the lowest tests, and work out the average modulus of elasticity for the same beams, we find that the beams of lowest strength have, as a rule, the lowest resistance to deflection, but that the weak beams have a much greater *proportionate* resistance to deflection than the strong beams.

[1] Roughly as 7 is to 4.

This, according to Professor Lanza's tests, is true for woods of very dissimilar character. Bauschinger's formula gives diametrically opposite results, and cannot therefore be accepted even if the constants are varied for different kinds of timber. Table XV contains a summary of the highest and lowest averages worked out from Professor Lanza's tests, with a

TABLE XV.—RELATION BETWEEN COEFFICIENT OF RUPTURE (f) AND MODULUS OF ELASTICITY (E)

No.	Kind of Timber.	Cross-breaking Tests.		Average Value of E for the same Beams.	Value of E as calculated from f by the Author's Formula
		Classification.	Average Value of f.		
1.	White Oak	Five highest	8,039	1,328,032	1,300,070
	„ „	„ lowest	4,112	909,318	929,740
2.	Hemlock	„ highest	5,890	1,211,160	1,204,818
	„	„ lowest	2,137	711,496	725,654
3.	Yellow Pine	Ten highest	10,548	2,219,624	2,156,700
	„ „	„ lowest	4,476	1,379,920	1,404,900
4.	White Pine	Five highest	6,244	1,319,762	1,319,467
	„ „	„ lowest	2,686	886,160	865,394
5.	Spruce	Ten highest	7,662	1,717,220	1,750,600
	„	„ lowest	2,351	987,378	969,600

column added to show the values of E which would be obtained from the values of f by means of a new formula devised by the writer, namely—

$$E = C\sqrt{f} \qquad (36)$$

Where E = modulus of transverse elasticity,

f = coefficient of transverse rupture,

and C = a constant varying for the different materials, as follows:—White oak, 14,500; hemlock, 15,700; yellow pine, 21,000; white pine, 16,700; spruce, 20,000.

These tests show that the value of E for each kind of timber varies almost exactly as $\sqrt{f}$. In other words, a weak specimen is relatively stiffer than a strong one; thus, the strength ratio between the five weakest oak beams and the five strongest is as 1 to 2, but the stiffness as $1\frac{2}{5}$ to 2.

CHAPTER V

FACTORS OF SAFETY

It is sound practice to make every part of a structure stronger than is absolutely necessary, but, in purely utilitarian structures, this excess of strength must not be so great as to entail extravagant waste of material. The ordinary method of procedure is to design each piece of timber so that the working load upon it does not exceed a certain fraction of the breaking weight. Thus, if a beam is calculated to break under a load of 20,000 lbs., the safe working load will be assumed to be one-tenth or one-eighth or some other fraction of 20,000 lbs.

Much uncertainty exists as to the proper factor of safety to be employed in timber structures. In this country one-tenth of the breaking weight has been largely adopted, but during recent years numerous tests of larger scantlings of unselected material have been made, with the result that the coefficients of strength have been considerably reduced, so that in many cases a structure designed in accordance with modern theory with a factor of safety of 5 will be as strong as a structure designed according to earlier coefficients with a factor of safety of 10. The selection of a proper factor of safety will therefore depend largely upon the nature of the coefficients which are adopted in the calculations. Other conditions, however, must also be taken into consideration, and much must be left to the judgment and experience of the designer.

Temporary structures, of course, do not call for as large a factor of safety as permanent structures. In calculating the deflection of a beam as large a factor of safety need not be employed as in calculating its transverse strength; in the one case overloading would involve nothing worse than a little increase in the deflection, while in the other it might involve the collapse of the structure. Again, the strength of timber is less variable under certain stresses than under others, and the more variable the behaviour the greater must be the factor of safety.

The report of a committee of the International Association of Railway Superintendents of Bridges and Buildings—a North American association—contains a valuable passage on this subject. As the coefficients of strength recommended by this committee have been given in Tables VIII and XII, it is interesting to know the factors of safety which, in the opinion of the committee, ought to be adopted with these coefficients, and we will therefore quote the passage in full:—

"It is difficult to give specific rules. The following are some of the controlling questions to be considered:—

"The class of structure, whether temporary or permanent, and the nature of the loading, whether dead or live; if live, then whether the application of the load is accompanied by severe dynamic shocks and pounding of the structure. Whether the assumed loading for calculations is the absolute maximum, rarely to be applied in practice, or a possibility which may frequently take place. Prolonged, heavy, steady loading, and also alternate tensile and compressive stresses in the same place, will call for lower averages. Information as to whether the assumed breaking stresses are based on full-size or small-size tests, or only on interpolated values averaged from tests of similar species of timber, is valuable in order to attribute the proper degree of importance to recommended average values. The class of timber to be used, and its condition and quality. Finally, the particular kind of strain the stick is to be subjected to, and its position in the structure with regard to its importance and the possible damage that might be caused by its failure.

"In order to present something definite on this subject, the committee presents the accompanying table, showing the average safe allowable working-unit stresses for the principal bridge and trestle timbers, prepared to meet the average conditions existing in railroad timber structures, the units being based upon the ultimate breaking-unit stresses recommended by your committee, and the following factors of safety, viz.:—

Tension with and across grain,	10
Compression with grain,	5
Compression across grain,	4
Transverse rupture, extreme fibre stress,	6
Transverse rupture, modulus of elasticity,	2
Shearing with and across grain,	4."

The following statements of Mr. B. E. Fernow, Chief of the U.S. Division of Forestry, will be of service in connection with the tests made under his direction, and recorded in the various tables of this section:—

"As to factors of safety, it may be proper to state that the final aims of the present investigations may be summed up in one proposition, namely, to establish rational factors of safety. It will be admitted by all engineers that the factors of safety as used at present can hardly be claimed to be more than guesswork. There is not an engineer who could give account as to the basis upon which numerically the factors of safety for wood have been established as '8 for steady stress; 10 for varying stress; 15 for shocks' (see Merriman's

Text-book on the Mechanics of Materials); or as 4 to 5 for 'dead' load and 5 to 10 for 'live' load (see Rankine's *Handbook of Civil Engineering*).

"The directions for using these indeterminate factors of safety given in the text-books would imply that the student or engineer is, after all, to rely on his judgment as to the modification of the factor, *i.e.* he is to add to this general guess his own particular guess. The factor of safety is in the main an expression of ignorance or lack of confidence in the reliability of values of strength, upon which the designing proceeds, together with an absence of data upon which to inspect the material. With a larger number of well-conducted tests, coupled with a knowledge of the quantitative as well as qualitative influences of various factors upon strength, and with definite data of inspection which allow ready sorting of material, the factor of safety, as far as it denotes the residuum of ignorance which may be assumed to remain, as to the character and behaviour of the material, may be reduced to a minimum, restricting itself mainly to the consideration of the indeterminable variation in the actual and legitimate application of load.

"While the values given in these tables may claim to contain more elements of reliability than most of those published hitherto, much more work will have to be done before the above-stated aim will be satisfied."

Professor Lanza is of opinion that, if the values of breaking strength obtained from large scantlings are used in the calculations, "a factor of safety 4 will be sufficient for most ordinary timber constructions", but that, in "mill-work and in other cases where there is the jarring of moving machinery, it is advisable to use a somewhat larger factor". These factors are intended only for timber posts, and for beams designed with reference to their breaking weight. For the modulus of elasticity he suggests that the average of quick laboratory tests ought to be divided by two, in order to allow for the increased deflection caused by long-continued loads; this would be equivalent to working the beam up to its full calculated resistance. The values of strength, according to Professor Lanza, ought to be those obtained by testing unseasoned scantlings, as it would seem, he says, "unless future experiments shall prove the contrary to be true, that we cannot rely, in our constructions, upon having any greater strength than that of the green lumber".

Thurston, in his *Materials of Engineering*, recommends for timber a "factor of safety of at least 5 under absolutely static loads, and when the uncertainties of ordinary practice as to the exact character of material, and especially where shake and the impact of live loads were to be considered, would make the factor not less than 8, and for much of our ordinary work 10".

If the values of breaking strength are the averages of a large number of tests, and if the values obtained by different experimentalists are in close accord, a lower factor of safety may be used than if the values are obtained from a few tests and are not corroborated by other experimentalists. For this reason the tables of breaking strength which have been given in this section include tests by a number of different persons of different nationalities. The reader can therefore select his own coefficients, and can adopt a high or low factor of safety according as the coefficient seems trustworthy or otherwise. The collapse of the grand stand at the Glasgow football-ground in 1902 shows that timber structures intended for live loads ought to have an ample margin of safety, and this is particularly necessary in structures exposed to the weather.

Section VII—CARPENTRY

BY

THE EDITOR

Section VII—CARPENTRY

CHAPTER I

JOINTS

Carpentry is the art of combining pieces of timber to support a weight, or to resist force. It is broadly distinguished from *joinery* by this, that while the work of the carpenter is essential to the stability of a structure, the work of the joiner is applied more to those details of construction and decoration which render the structure more pleasing to the eye and more comfortable.

It is a mistake to imagine that a lower standard of workmanship is involved in carpentry than in joinery. The framework on which the stability of a structure depends ought not to be scamped, even though it is to be hidden, perhaps, by boarding or plaster. But the work of the carpenter is not always concealed, or of a merely utilitarian character. Roof-trusses, for example, are often of elaborate design, involving skilled workmanship of a high order.

In designing a timber structure, the several pieces must be so arranged as to be in equilibrium under the severest stresses to which the structure will in practice be subjected. This involves a knowledge of statics and of the strength of timber, as explained in the previous section, and also of the best methods of connecting the pieces so that the stresses will be transmitted from piece to piece without yielding at the joints.

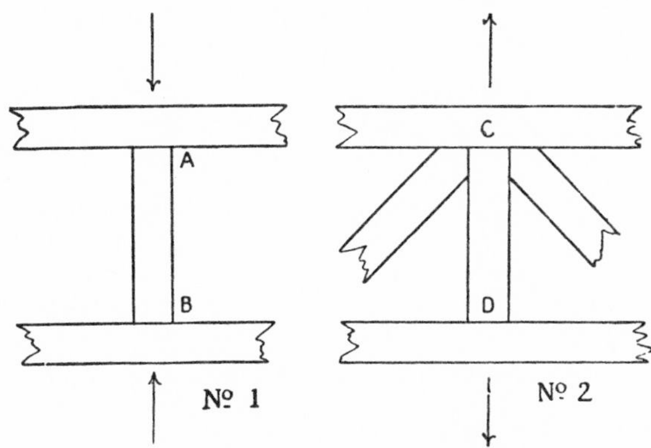

Fig. 579.—Members in Compression and Tension

We have already seen that, in a timber truss, some members may be in tension, others in compression, and others again may be under a transverse or bending stress. It is obvious that a joint which is suitable for the end of a compression member or strut, may be utterly useless for the end of a tension member or tie. Thus, in fig. 579, No. 1, the vertical member A B is in compression, and if the load is such as to give a uniformly vertical stress, the ends may be cut perfectly square to form simple butt joints at A and B. But if a vertical member forms part of a truss, as shown at C D in No. 2, it may be in tension, and a butt joint is obviously useless; what is required is a form of joint which will enable the pull in C D to be transmitted to the upper and lower horizontal members or chords. In other words, the vertical member must be rigidly fixed at C and D in such a manner as to prevent the horizontal members being pulled apart if forces are applied to them as shown by the arrows. In some cases a member may be in tension under one form of loading and in compression under another, and the joints must therefore be adapted both for tension and compression.

The form of the joints in carpentry is therefore governed by the nature of the stresses which the joints are required to resist or transmit. In this chapter some typical examples of the kinds of joint most commonly used will be given, so that the student will be able to grasp the principles governing the design of the joints described and illustrated in subsequent chapters.

COMPRESSION JOINTS.—These are not difficult to design. The principal points to be borne in mind are the following:—

1. To form the abutting surfaces as nearly as possible perpendicular to the line of action of the stress in the compression member.

2. To form the abutting surfaces with the greatest possible accuracy so that the pressure is equally distributed over the joint.

3. To proportion the area of the abutting surfaces according to the stress which the joint has to resist or transmit, allowing a proper margin of safety.

4. To provide against lateral movement of the members forming the joint.

5. To form the joint in as simple a manner as possible, and with the edges of the bearing surfaces exposed to view.

The simplest compression joint is formed with two flat surfaces, as in the case of a post supporting a beam, lateral movement being prevented by two or more wrought-iron "dogs" driven into the two members, as shown at A in fig. 580. This joint does not weaken the timber beyond the very slight weakening effect of the dog, which will tend to split the post somewhat before the ultimate breaking weight is reached. The iron dog is unsightly, and the joint is not often used for permanent structures, although it is certainly one of the best compression joints which can be made. Instead of iron dogs, drift-bolts are largely used in America, two for fastening the post to the sill (shown at B in fig. 580), and one for fastening it to the cap, the latter bolt being driven vertically downwards through the cap into the post. The hole through the first piece of timber pierced should be very nearly the same size as the bolt; the remainder of the hole should be slightly smaller. The disadvantage of the top drift-bolt is that it renders repairs difficult. A better method is to use wrought-iron straps (fig. 581, A).

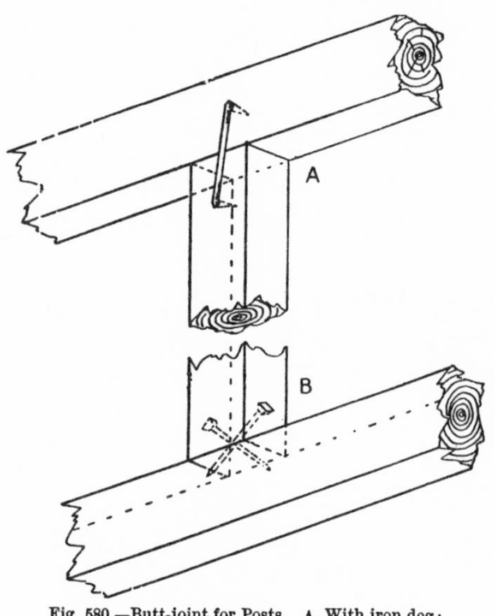

Fig. 580.—Butt-joint for Posts. A, With iron dog; B, with drift-bolts.

Instead of dogs or drift-bolts two dowels are often used to prevent lateral motion. These may be of $\frac{5}{8}$- or $\frac{3}{4}$-inch round or square wrought-iron, and about 5 inches long, for posts about 12 inches square.

Still another method of forming the joints between posts and beams, without wasting materially the length of the posts, is that known in America as the "plaster" joint, and in this country as the "fished" joint. The post is housed into the sill and cap about 1 inch deep, and each joint is secured by two cover-pieces, one on each side, spiked and bolted to the post and sill or cap (fig. 581, B). The joint is simple and strong, but the wood fish-plates or plasters are certainly not ornamental. Iron straps are neater, as shown at A; the entrance of moisture can to some extent be prevented by chamfering the top of the post. With these joints repairs are easily executed.

In this country lateral movement is often prevented by forming a tenon or joggle on the post and fitting it into a corresponding mortise in the beam or sill. It may be said, in passing, that in the case of a mortise in a sill exposed to the weather, a hole should be bored from the bottom of it outwards and downwards to the face of the sill, as water entering the joint will otherwise be unable to escape and will inevitably hasten the decay of the timber.

The mortise-and-tenon joint is perhaps the most common form of compression joint used in this country for timbers meeting at right angles, as in the case of sills or beams and posts. Indeed it is the most generally useful joint both for the carpenter and joiner. In the simplest case of a tenon-and-mortise joint, the two pieces of wood meet at right angles

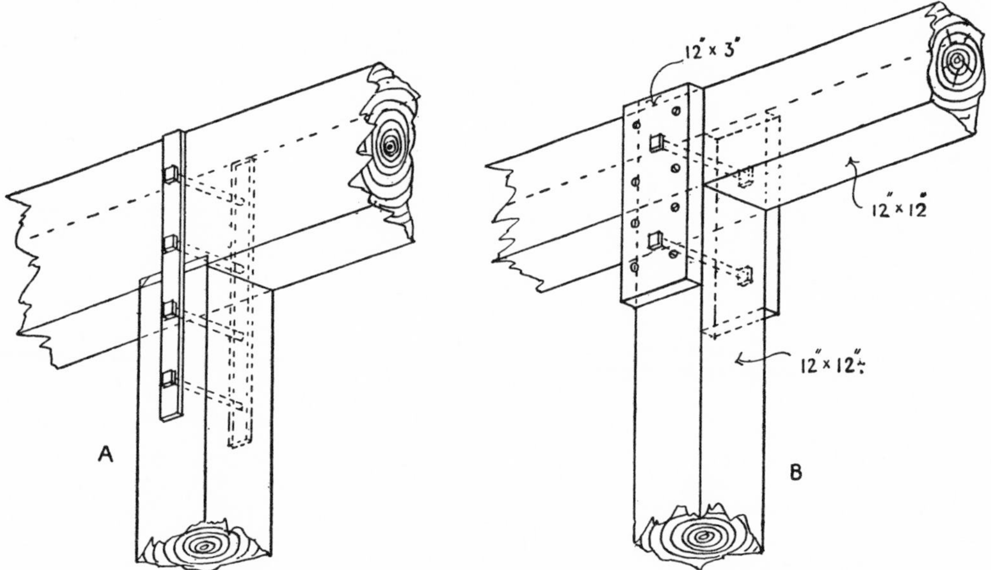

Fig. 581.—Joints between Posts and Beams. A, With iron straps and bolts; B, with wood fish-plates, and iron bolts and spikes.

(fig. 582). The tenon *a* is formed at the extremity of the piece A, in the direction of its fibres and parallel to its axis *m n*, by two notches, which take from each side a parallelepipedon. The planes of the sides *fg* of the tenon are parallel to the face *b* of the timber, and the other planes of the notch at right angles to it. The thickness *fg* is generally one-third of the thickness of the timber A.

The mortise is hollowed in the face of the piece B, and is of exactly the same length and breadth as the tenon, which therefore perfectly fills it. The two sides of the mortise which correspond to the breadth of the tenon should be parallel to the direction of the fibres of the wood. The sides of the mortise are called its *cheeks*, and the square parts of the timber A from which the tenon projects, and which rest on the cheeks of the mortise, are called the *shoulders* of the tenon, and its springing from these is called its *root*.

Fig. 582.—Mortise-and-Tenon Joint

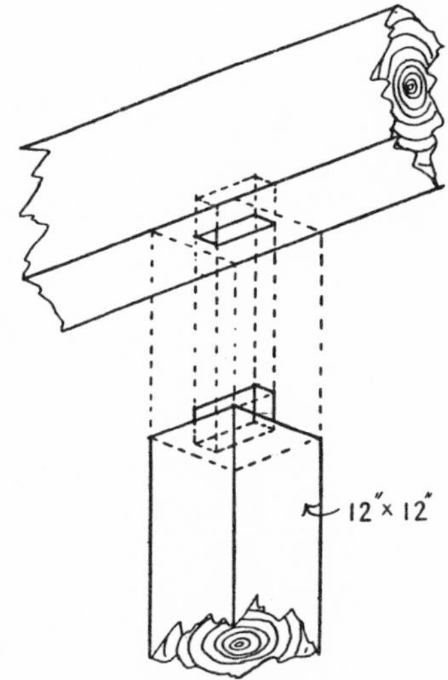

Fig. 583.—Mortise and Stump-tenon or Joggle

When the mortise-and-tenon joint is cut, adjusted, and put together, the pieces are often united by a key or trenail. The key is generally round, with a square head, and in diameter is about equal to a fourth part of the thickness of the tenon. It is generally inserted at the distance of one-third of the length of the tenon from the shoulder.

But a key should never be depended on as a means of securing the joint; for the immobility of a system of framing should result from the balancing of the forces and the precision of the execution. A frame fixed definitely in its place should be stable and solid without the aid of keys, which are to be regarded generally as mere auxiliaries, useful during its construction.

In fig. 583, shoulders are formed on all the four sides of the tenon; this gives a better bearing on the beam. Such a tenon is known as a "stump-tenon" or "joggle".

The mortise must be made slightly deeper than the length of the tenon, so that the shoulders of the post will come to a bearing against the beam. The load is therefore distributed only over the shoulders of the post, and this reduces the effective area of the bearing surfaces and increases the risk of the beam being indented. We have seen in Chapter III, Part II of the preceding section that a comparatively small load will produce a 3-per-cent indentation of timber when applied across the grain, and we have also seen that the unequal indentation, which occurs owing to the hard and soft portions of the grain, reduces considerably the ultimate resistance of the post through or to which the load is applied. For these reasons the tenon ought not to be made larger than necessary; the resistance of

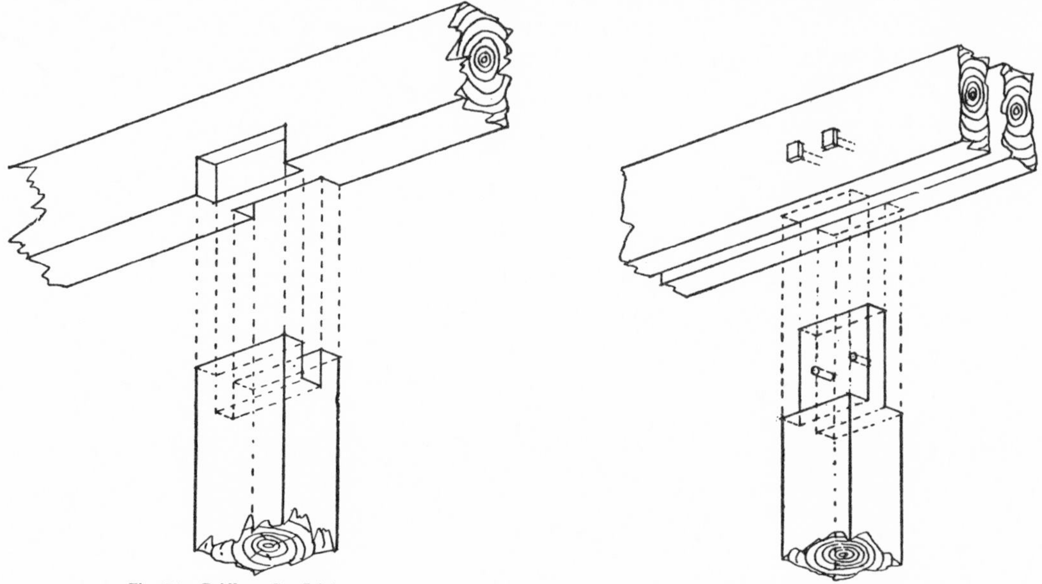

Fig. 584.—Bridle or Cog Joint

Fig. 585.—Joint between Post and Double Beam

timber to shearing across the grain is so great that there is little danger of the tenon failing in this manner. Again, the tenon requires, as a rule, only a small projection, and involves only a shallow mortise, which has the advantage of not reducing to a serious extent the substance of the beam. Sometimes, however, one or two hardwood pegs or trenails are used for securing the joints and drawing the timbers together; in such cases the tenon must be somewhat longer. In America the tenons on 12-inch square posts, to be secured with trenails, are generally about 8 inches by 3 inches, and 5 inches long, the trenails being 1 to $1\frac{1}{4}$ inch in diameter. The area of the tenon is therefore 24 square inches, or one-sixth of the sectional area of the post.

Sometimes two notches are cut in the beam, one on each side, leaving a bridle or cog between them to fit into a corresponding groove or sinking in the head of the post (fig. 584). This is not as good as the mortise-and-joggle joint, because there is greater difficulty in making the tops of the two notches absolutely level than in making the top of the post level around the tenon, and also because the load is transmitted to the two side portions only of the post, and not to all the four sides, as in the mortise-and-joggle joint.

Another method frequently used in American bridges consists in using two beams spaced about 3 inches apart, and forming two shoulders (with a tongue between) on the head of the post (fig. 585). Thus, instead of one beam 12 inches square, two beams 6 inches by 12 inches will be used, and the joint will be secured by one or two $\frac{3}{4}$-inch or 1-inch bolts.

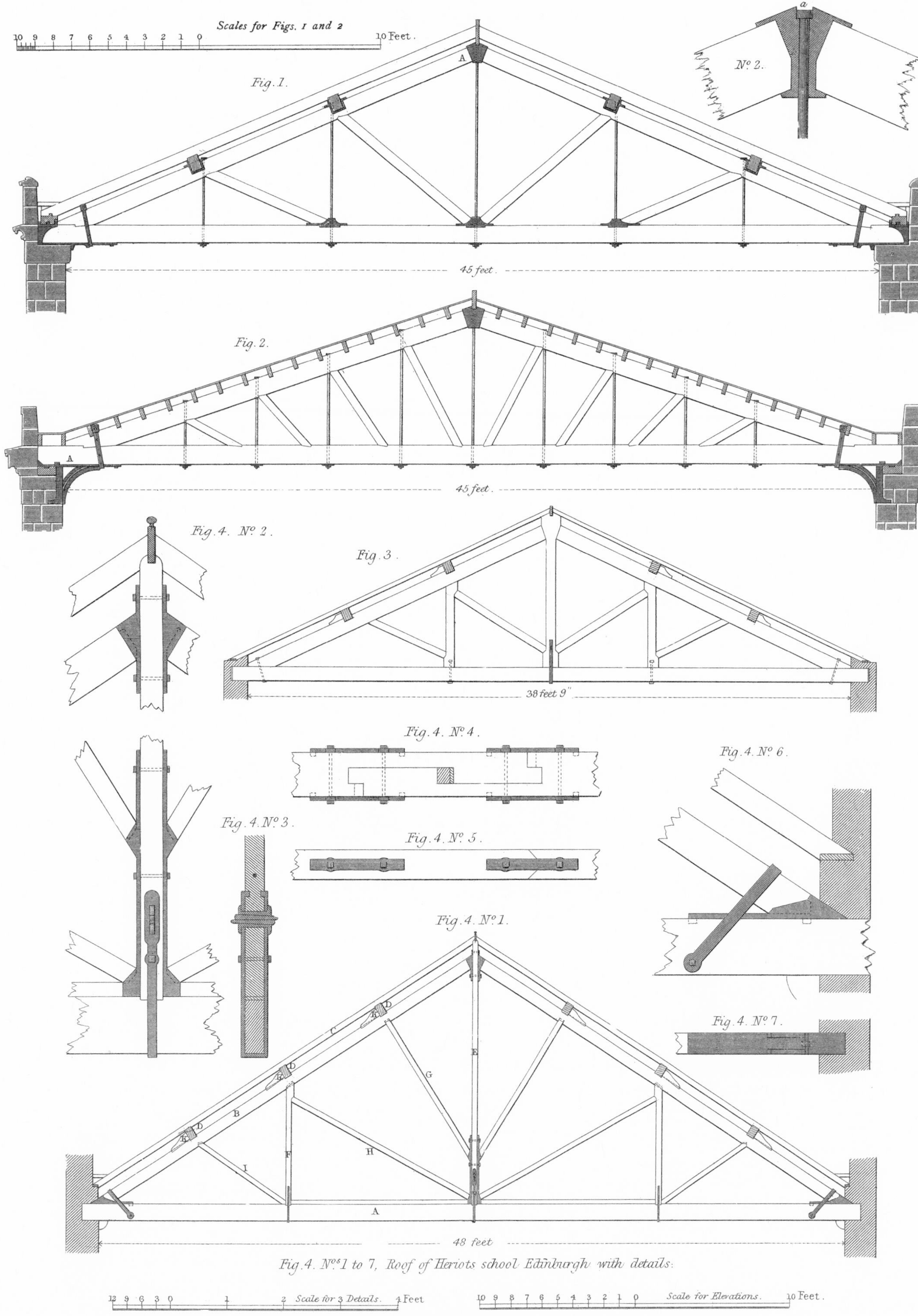

Fig. 4. Nos 1 to 7, Roof of Heriots school Edinburgh with details.

COMPOUND KING-POST TRUSSES

PLATE XXVII

COMPOUND KING-POST TRUSSES

Fig. 1.—Elevation of Compound King-post Truss with wrought-iron suspension-rods and timber tie-beams, principal rafters, and struts; the two purlins on each side rest in cast-iron chairs; No. 2 is an enlarged drawing of the cast-iron socket at A.

Fig. 2.—Elevation of Compound King-post Truss with wrought-iron suspension-rods, and with the roof-boarding nailed directly to the purlins, which are spaced at short intervals for the purpose.

Fig. 3.—Elevation of Compound King-post Truss, all the members being of timber.

Fig. 4.—No. 1: Elevation of Compound King-post Truss, George Heriot's School, Broughton Street, Edinburgh.

Nos. 2 and 3: Details of King-post.

Nos. 4 and 5: Details of Scarfing of Purlins.

Nos. 6 and 7: Details of End of Tie-beam.

These illustrations are more fully described on pages 38–40, Volume II (Divisional-Volume V).

For beams of this size the tongue on the post would be about 3 inches thick. The advantages claimed for this method of construction are—that the timber required for the beams is smaller, and therefore cheaper and easier to handle; that new beams can be more easily substituted; that in many cases only one of the two will require to be replaced, whereas if only one beam had been used, this beam of double the size would have had to be replaced; and that the beams can be removed without damaging any other part of the structure. A further advantage is that the tenon or tongue is well ventilated, and therefore less likely to decay.

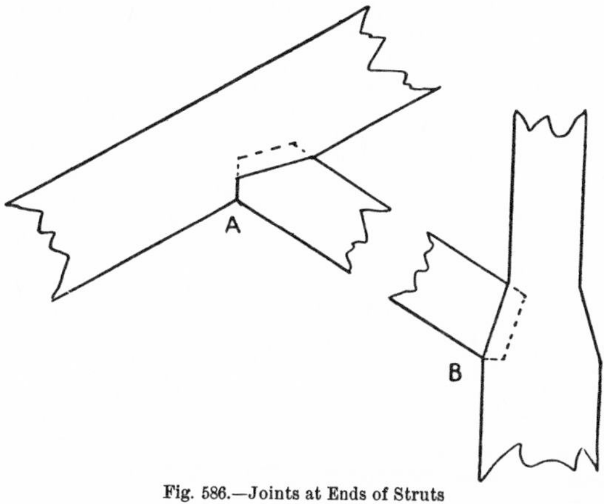

Fig. 586.—Joints at Ends of Struts

Occasionally the top of a post is cut to the shape of a wide inverted V, and fitted into a corresponding notch in the beam; but this method cannot be recommended, nor can the modification of it in which the head of the post is cut to a segmental form, as it is almost impossible to make the abutting surfaces fit against each other accurately, and the strength of the post is consequently reduced.

Slight modifications in the form of the compression joint are made when the two members do not meet at right angles. The commonest example of this is the joint formed by the strut and rafter of a roof-truss, the strut being housed and tenoned into the rafter, as shown in fig. 586 at A. The joint between the foot of the strut and the king-post or queen-post (shown at B) is very similar.

TENSION JOINTS.—The best examples of tension joints are those required for lengthening the tension members of timber framing, such as the tie-beams of roof-trusses and the lower chords of bridge-trusses, but these will be considered in the next chapter.

In the preceding section it was shown that timber has a relatively great tensile strength, but that in practice this strength cannot be utilized to the full on account of the difficulty of fixing the ends. A tension member in a timber structure will almost invariably give way by shearing at the joint long before the ultimate tensile strength of the member itself is reached. For this reason the strength of tension joints is generally eked out by means of wrought-iron fastenings of various kinds.

The most common examples of tension joints are those at the head and foot of king- and queen-posts in roof-trusses, and at the ends of tie-beams. The joint at the foot of a king-post or queen-post may be taken as a typical example. The common form of these joints is shown in fig. 587, No. 1. The king-post B, if it fulfils its duty in the truss, is in tension, and must therefore be so fixed that it can pull against the tie-beam A. To prevent lateral movement the king-post is tenoned into the tie-beam, but it is obvious that such a joint is absolutely useless for transmitting the tensile stress in B to A. Some kind of fastening must be adopted to make the joint secure. This sometimes takes the form of a bolt passing up through the tie-beam into the king-post to a point a little below or above *ff*, a hole being sunk in the king-post at this point, so that a nut and washer can be placed on the end of the bolt and the bolt be drawn tight. The more common method is shown in the illustration, and consists in the use of a wrought-iron strap or stirrup *a b* wedged tight by means of the gib-and-cotter joint at *c*. The bolt in the one case, and the strap in the other, must be strong enough to bear safely the whole of the stress in the king-post.

No. 2 shows a form of joint used when an iron king-rod takes the place of the timber king-post. A cast-iron socket-piece is bolted to the tie-beam, having a central hole through which the king-rod passes, and sockets for the ends of the two struts BB. The metal socket-piece is often omitted, the two struts being housed and tenoned into the tie-beam, with or without a short straining-piece between.

No. 3 shows the joint commonly used for the foot of the queen-post, A being the tie-beam, B the queen-post, C the strut, and D the straining-piece.

In Hurst's edition of Tredgold's *Carpentry*, the strength of straps for roofs is given as follows:—Where the longest unsupported part of the tie-beam is

10 feet,	the strap may be	1 inch wide	by $\frac{3}{16}$	inch thick
15 „	„	$1\frac{1}{2}$ „	„ $\frac{1}{4}$	„ „
20 „	„	2 inches	„ $\frac{1}{4}$	„ „

These dimensions are said to be sufficient for ordinary roofs, but if the tie-beam is transversely loaded, the strength of the straps must be proportionately increased.

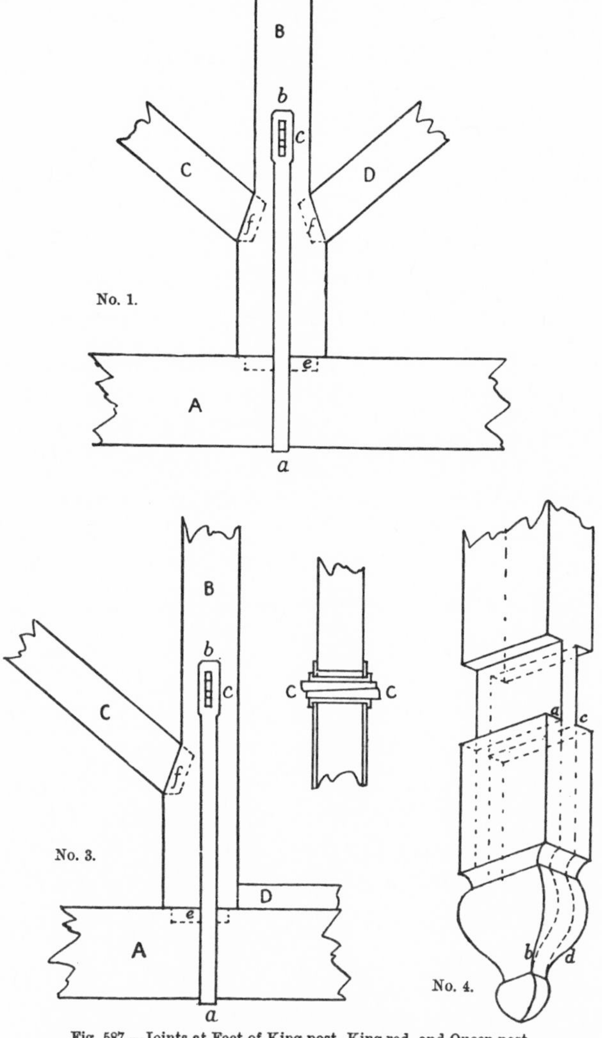

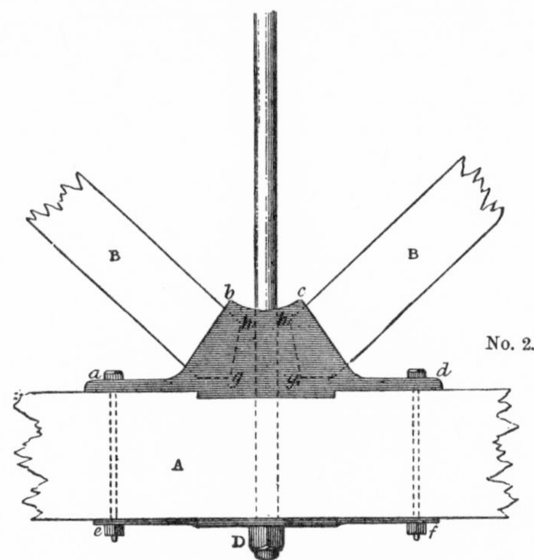

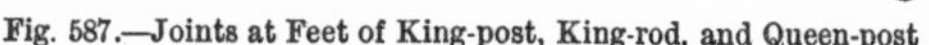
Fig. 587.—Joints at Feet of King-post, King-rod, and Queen-post

If the joint must be made without metal fastenings, an entirely different arrangement is necessary. A long tenon might be formed on the king-post and passed through a corresponding mortise in the tie-beam, and secured with a wood wedge passing through a hole in the tenon immediately below the tie-beam, or with wood trenails driven through the tie-beam and tenon, but the resistance of such a joint would be comparatively small. A better method is to make the tie-beam in two pieces, to prolong the king-post some distance below the tie-beam, and to cut a notch on each side of it to receive the two halves of the tie-beam (fig. 587, No. 4). This arrangement has been adopted in practice, and has proved satisfactory. The pieces must be fastened together by trenails, spikes, or bolts. In the illustration only the king-post is shown; the two halves of the tie-beam are not cut in any way, but merely housed into the notches of the king-post. The strength of this joint will depend partly upon the strength of the trenails, but chiefly upon the tensile resistance of the reduced cross-section of the king-post, and upon the area of the two planes *a b* and *c d*, along which shearing may occur below the notches. The longer this extension of the tie-beam is, the greater will be the area of these planes, and consequently the greater will be the shearing resistance.

A third factor, however, must be taken into consideration, namely, the resistance of the lower shoulders of the notches to compression.

An example of this kind of joint will be worked out, and will serve to show the methods of calculation which must be adopted in designing joints in framed structures. We

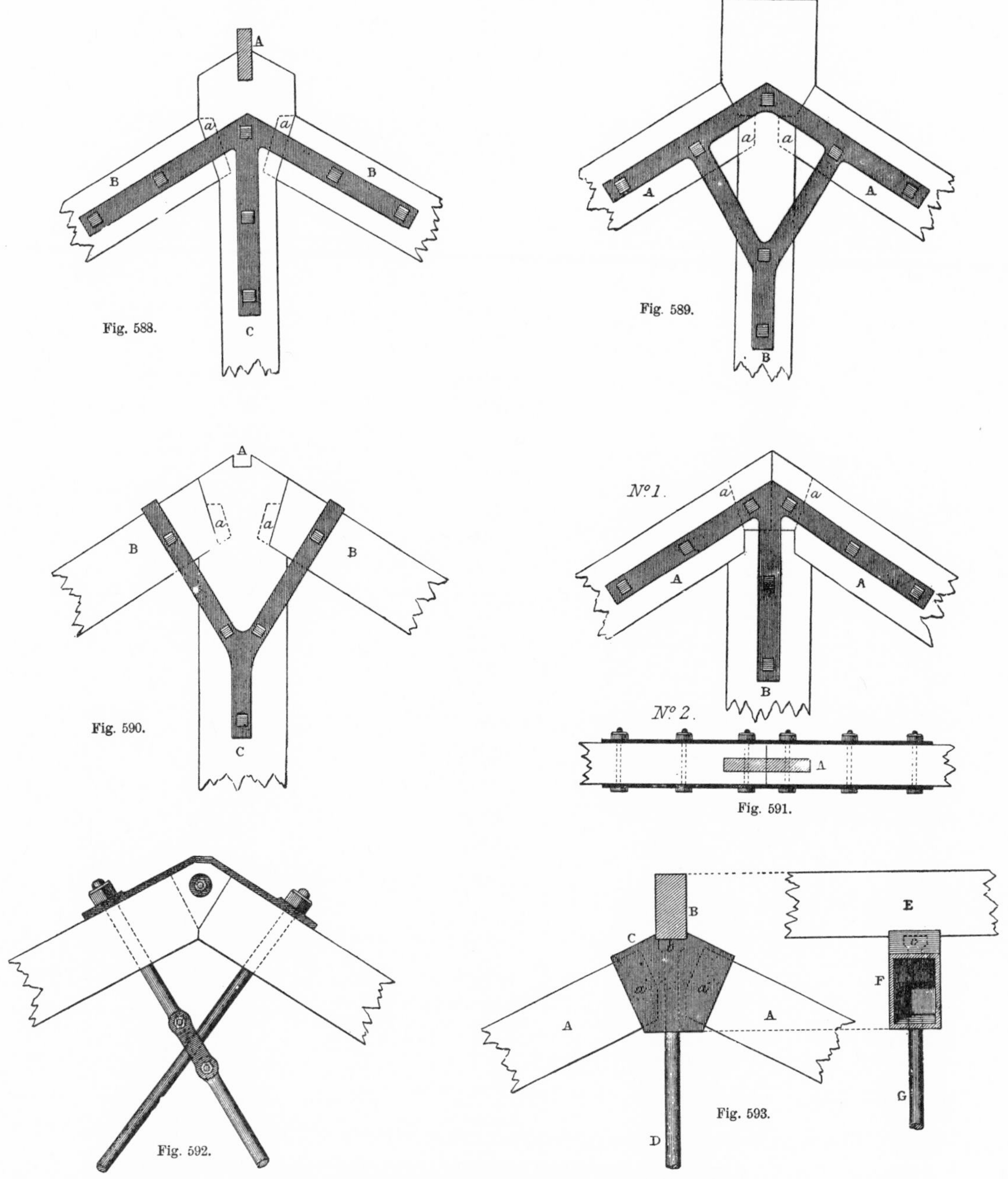

Figs. 588 to 593.—Joints at the Head of King-posts and King-rods

will assume that the truss is of pitch-pine and that the tensile stress in the king-post is 7500 lbs., and that the working resistances of pitch-pine are as follows:—

Tension	1500 lbs. per square inch.
Compression	1000 „
Shearing	150 „

From these figures the necessary areas to resist the three stresses can be computed as follows:—

$$\text{Tension area} = \frac{7500}{1500} = 5 \text{ square inches.}$$

$$\text{Compression area} = \frac{7500}{1000} = 7\tfrac{1}{2} \quad \text{''}$$

$$\text{Shearing area} = \frac{7500}{150} = 50 \quad \text{''}$$

To the net tension area must be added the hole for the trenail, say 1 square inch. The gross area must therefore be $5 + 1 = 6$ square inches. This will be obtained by a king-post

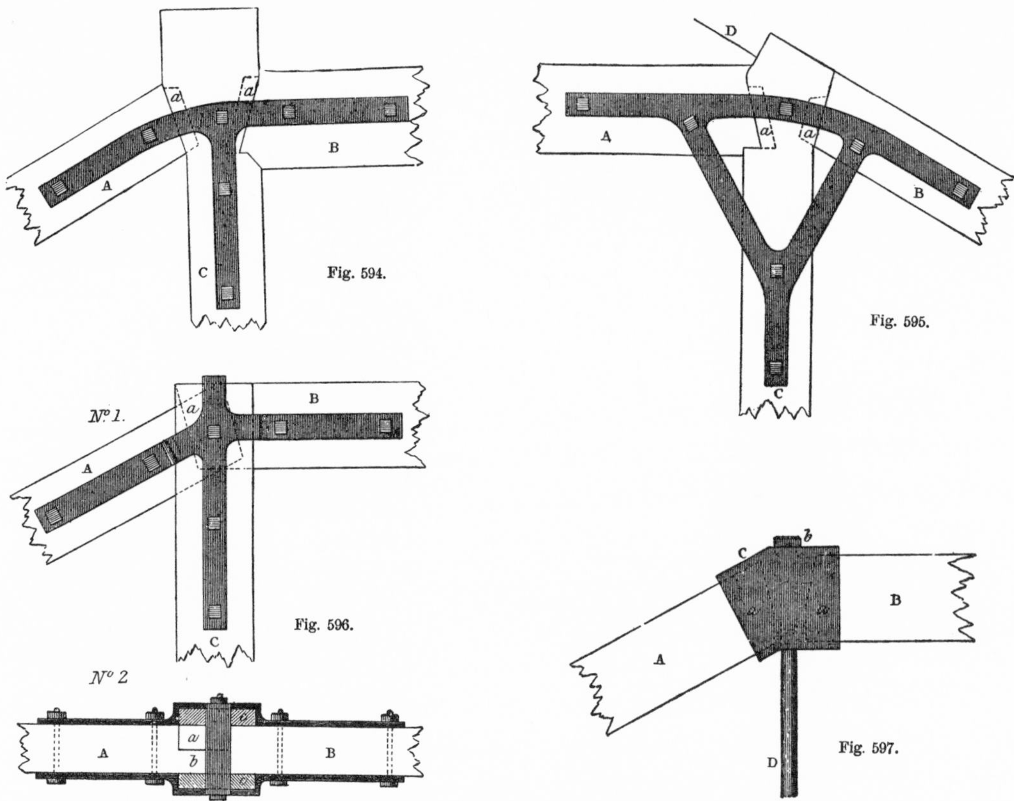

Figs. 594 to 597.—Joints at the Head of Queen-posts and Queen-rods

5 inches wide and $1\frac{1}{5}$ inch thick at the reduced part. The two compression areas must contain $3\frac{3}{4}$ square inches each, which will be obtained by two shoulders 5 inches × $\frac{3}{4}$ inch. The two shearing areas must contain 25 square inches each, that is to say, 5 inches × 5 inches. In this calculation the strength of the trenail is not considered. The scantling of the uncut king-post must therefore be 5 inches by $(1\frac{1}{5} + \frac{3}{4} + \frac{3}{4} =)\ 2\frac{7}{10}$ inches, or, say, 5 inches × 3 inches. In practice the compression areas should be somewhat larger, as the tie-beam is compressed *across* the grain by the shoulders of the king-post, and timber compressed in this way is indented under a comparatively low stress.

The joint at the head of a king- or queen-post is similar in principle, although it differs in detail, as the members forming the joint do not meet at right angles, and as the post is the only tension member, the others being in compression. Such joints may be known as tension-compression joints, but as there is little difficulty in forming the ends of the compression members, the problem really becomes one of tension or shearing. The common form of the king-post joint is shown in fig. 588, B B being the two rafters, tenoned at *a a* into the king-post C to prevent lateral movement. The shoulders of the tenons must be accurately fitted against the seats formed on the king-post, in order that the stresses may be equally distributed. The strength of the joint will depend upon the shearing resistance of the timber above the heads of the tenons *a a*, unless wrought-iron straps bolted on each side of the joint are used. Other forms of joints and straps for the head of a king-post are shown in figs. 589 to 591. The last is unnecessarily complicated. In fig. 592 two wrought-iron king-rods take the place of the king-post, one of them having a link to permit the passage of the other; this joint is seldom or never used in practice. Fig. 593 is a more common form; a wrought-iron king-rod is used instead of a timber king-post, and passes through an iron casting containing sockets for the ends of the rafters.

The corresponding joints for a queen-post truss are shown in figs. 594 to 597. In fig. 596 the rafter A and straining beam B are halved together, and the queen-post is formed of

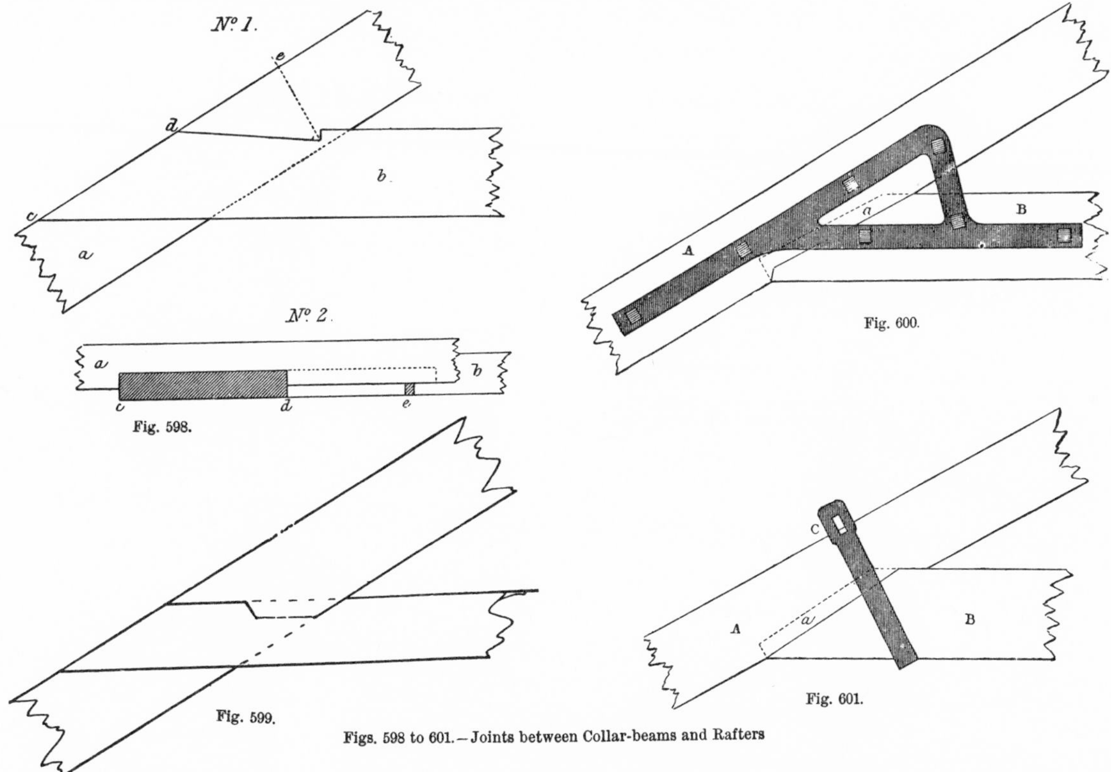

Fig. 598. Fig. 599. Fig. 600. Fig. 601.

Figs. 598 to 601.—Joints between Collar-beams and Rafters

two pieces placed one on each side of the joint A B; the four timbers are secured by straps and bolts, the straps being too intricate to be recommended. In fig. 597 a wrought-iron queen-rod D takes the place of the queen-post, and a cast-iron socket-piece receives the ends of the rod, rafter, and straining beam.

For the junction of collar-beams and rafters a dovetailed joint is often used, as shown in fig. 598. The collar (unless the outward thrust is counteracted by suitable abutments) is in tension, and the dovetail is intended to give it an anchorage to the rafter, but such a joint is almost certain to yield in consequence of the shrinkage in the depth of the collar. In the illustration the collar is 8 inches deep, the shoulder of the dovetail 1 inch, and the length of the dovetail 13 inches. If the collar shrinks $\frac{1}{12}$ inch in the depth—which is not by any means an abnormal amount—the rafters will spread 1 inch at each end of the collar before this comes to its final bearing; assuming the collar to be fixed at half the height of the rafters, this gives a spread of 4 inches across the foot of the truss. The joint can be improved by making the bearing surface of the dovetail or notch of quicker pitch, as shown in fig. 599, but in any case the joint ought to be further secured by trenails, spikes, bolts, or straps. Two forms of joint secured with straps are shown in figs. 600 and 601; in these the collar B is tenoned into the rafter A to prevent lateral movement. The strap or stirrup in fig. 601 must be drawn tight by wedges at C; this is not a good arrangement, as the direction of the strap is very different from the line of action of the stress in B. Sometimes a bolt is used instead of a stirrup, but in a similar direction.

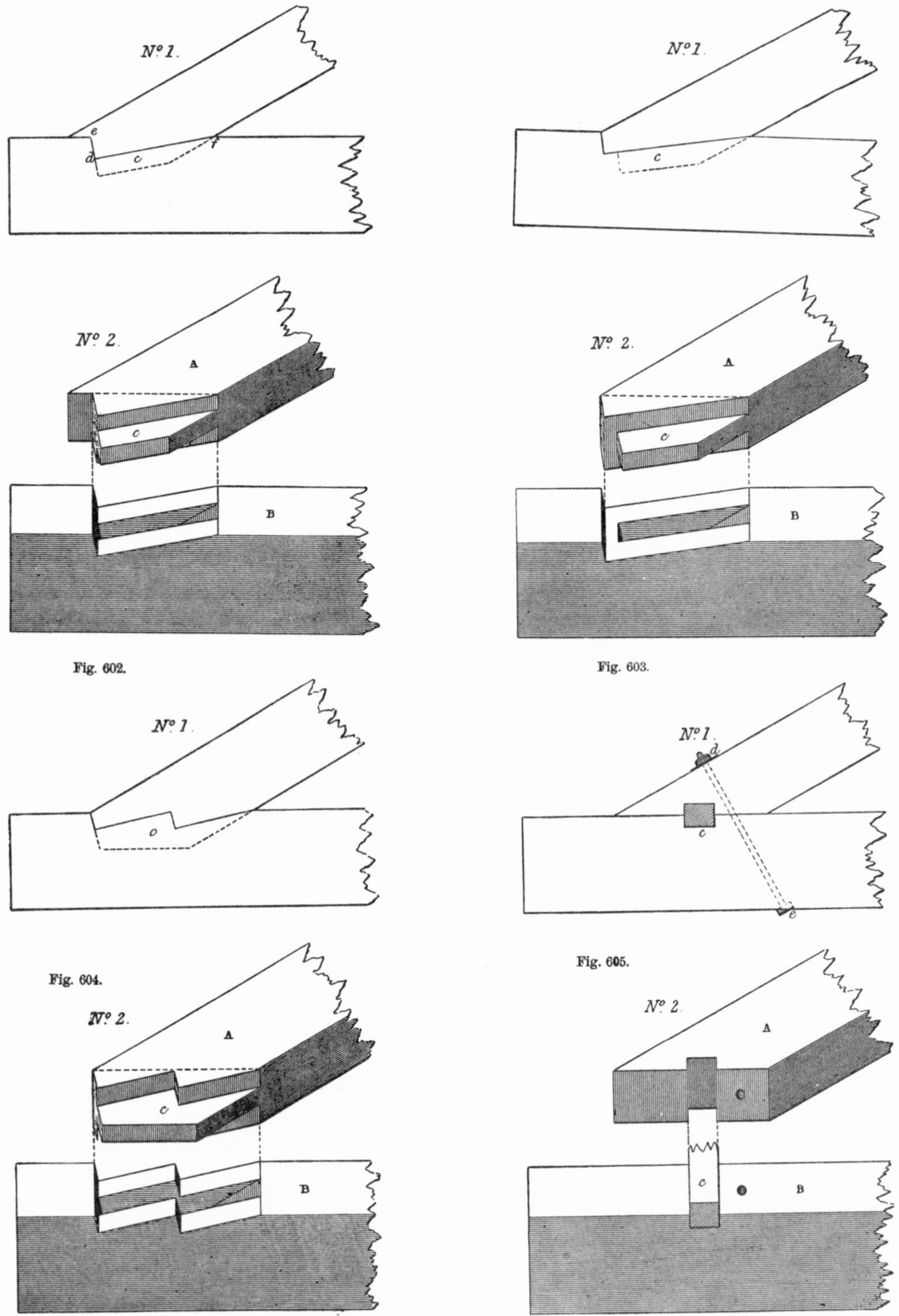

Figs. 602 to 605.—Joints between Principal Rafters and Tie-beams

The joints formed by the junction of principal rafters and tie-beams are, like those at the heads of king- and queen-posts, tension-compression joints, and must be formed so that the outward thrust in the rafters will be resisted by the tie-beams. Here again the strength of the joint, if iron fastenings are not used, will depend chiefly upon the shearing resistance which the end of the tie-beam beyond the joint can be made to offer. Different forms of the joint are shown in figs. 602 to 607.

Fig. 602.—No. 1 shows the joint in elevation, *c* being the tenon, truncated at the outer end. The cheeks of the mortise are cut down to the line *d f*, so that an abutment *e d* is formed of the whole width of the cheeks, in addition to that of the tenon; a *bird's-mouth* is formed on the end of the rafter to fit the notch and the top of the tie-beam. No. 2 shows the parts detached and in perspective.

Fig. 603.—No. 1 is a geometrical elevation of a joint, differing from the last by having the end of the rafter truncated, and the shoulder of the tenon returned in front. It is represented in perspective in No. 2.

Fig. 604.—Nos. 1 and 2 show the geometrical elevation and perspective representation of an oblique joint, in which a double abutment is obtained. In all these joints, the

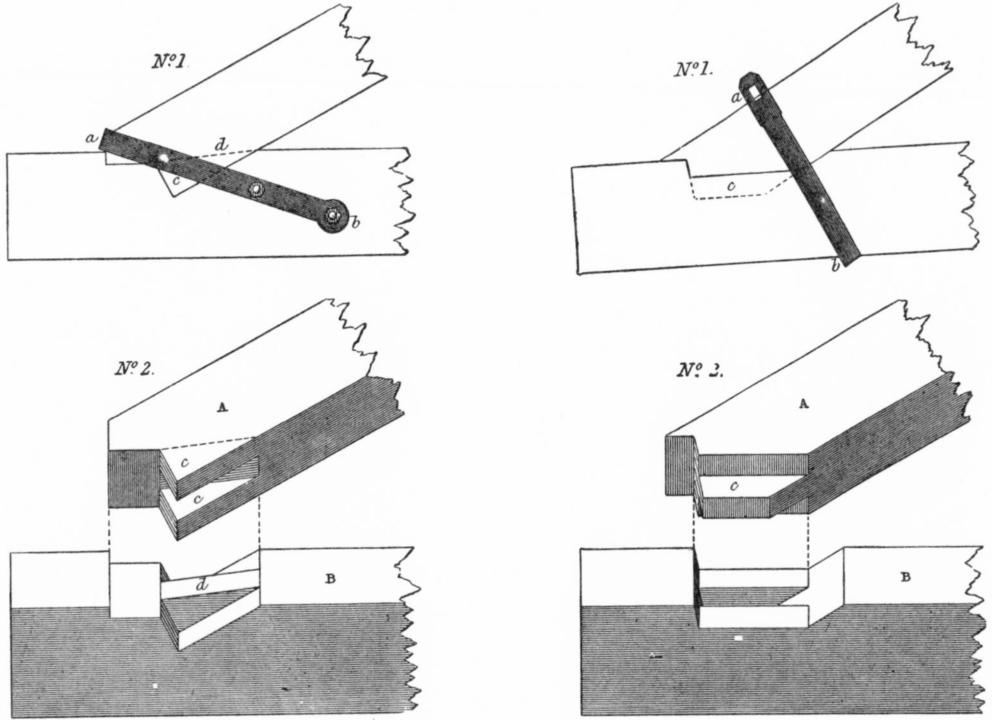

Figs. 606 and 607.—Joints between Principal Rafters and Tie-beams

abutment, as *d e*, fig. 602, should be perpendicular to the line *d f*; and in execution, the joint should be a little free at *f*, in order that it may not be thrown out at *d* by the settling of the framing. The double abutment is a questionable advantage; it increases the difficulty of execution, and, of course, the evils resulting from bad fitting. It is properly allowable only where the angle of the meeting of the timbers is very acute.

Fig. 605.—Nos. 1 and 2 show a means of obtaining resistance to sliding by inserting the piece *c* in notches formed in the rafter and the tie-beam; *d e* shows the mode of securing the joint by a bolt.

Fig. 606.—Nos. 1 and 2 show a form of joint in which the place of the mortise is supplied by a groove in the rafter, and the place of the tenon by a tongue or bridle *d* in the tie-beam. As the parts can be all seen, they can be more accurately fitted, which is an advantage in heavy work. In No. 1 the mode of securing the joint by a strap *a b* and bolts is shown.

Fig. 607.—This is another mortise joint, secured by a strap *a b* and cotter or wedge *a*. A bolt, as at *d e* in fig. 605, is often used instead of the strap shown in figs. 606 and 607.

Transverse-stress Joints, &c.—The name "transverse-stress joints" may be applied to the joints between beams and their bearers, whether the latter are other beams, wall-plates,

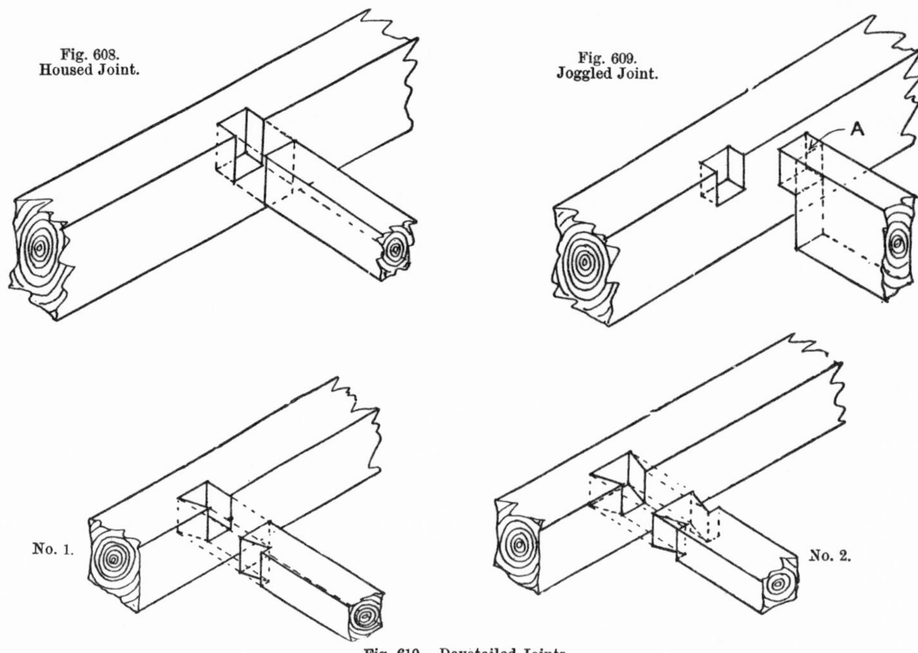

Fig. 610.—Dovetailed Joints

or posts, and also to the joints used for lengthening beams. The latter, however, will be considered in the next chapter.

The simplest form of joint is where the beam merely rests upon the support. Another simple form is where the beam is *housed* into the other member, as shown in fig. 608. This is not calculated to resist tension unless some kind of fastening is used. Sometimes the end of one piece is partly cut away to form a *joggle* (fig. 609); the projection A can be spiked to the beam or plate. A modification of fig. 608 is the *dovetailed joint* (fig. 610); this, if tightly made, affords some resistance to tension, although the amount of this resistance is a matter of uncertainty on account of the probable shrinkage of the dovetail. In No. 1 the dovetail is single, in No. 2 double. Dovetailed tenon joints secured with wedges *c* are shown in fig. 611, and a cogged tenon joint, which also requires a wedge, is shown in fig. 612. The mortise in the latter is difficult to make, and the joint cannot be recommended.

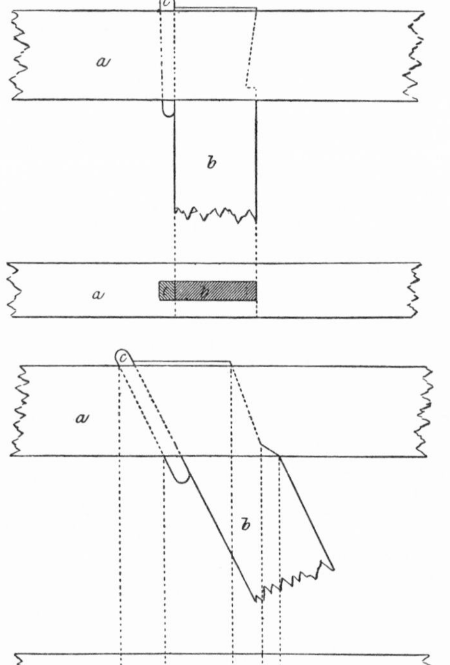

Fig. 611.—Dovetailed Tenon Joints secured with Wedges

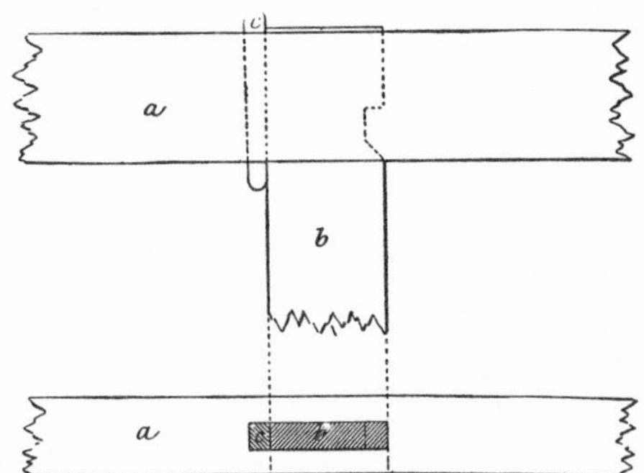

Fig. 612.—Cogged Tenon Joint secured with Wedge.

Notching (fig. 613) is often used for joists resting on wall-plates, and *cogging* (fig. 614) for these and many other joints; the cog may be in the centre of the upper surface of the lower piece, or, better, adjoining the inner edge; in the latter case only one notch has to be cut in the lower

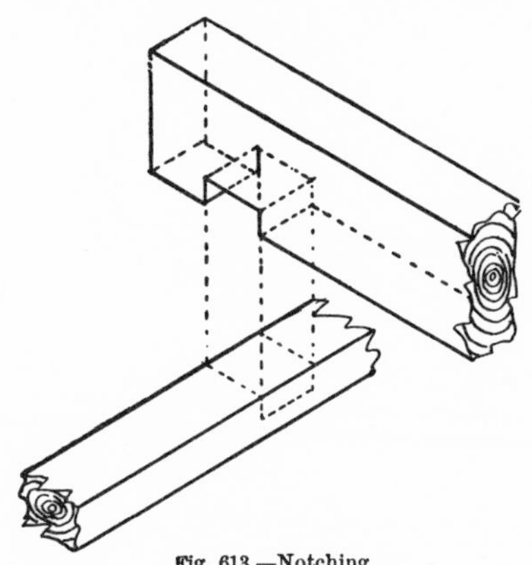

Fig. 613.—Notching

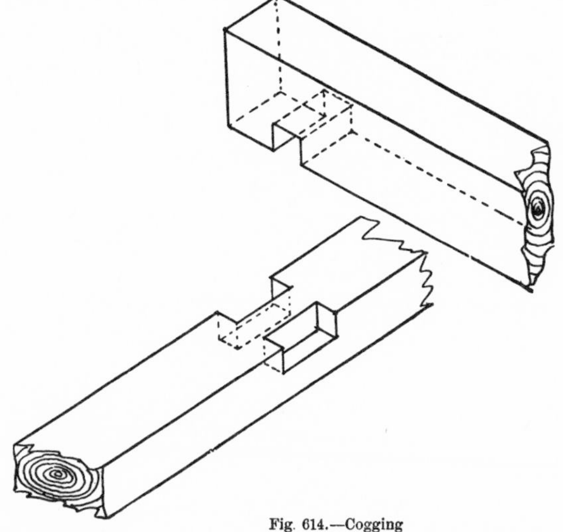

Fig. 614.—Cogging

piece instead of two, and a greater shearing resistance is offered by the end of the beam.

Halving may be either plain, bevelled, or dovetailed, as shown in fig. 615 at A, B, and C

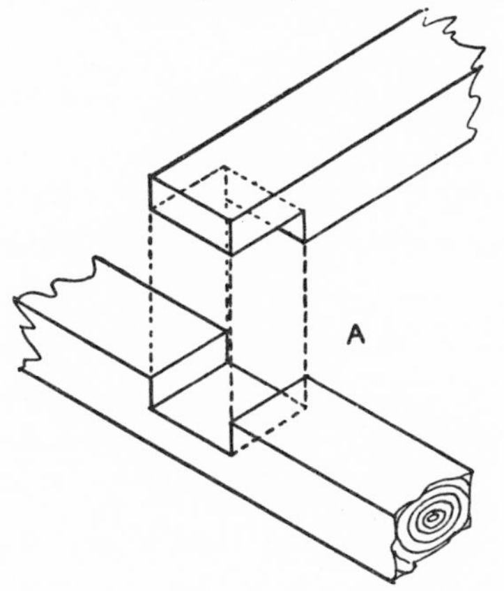

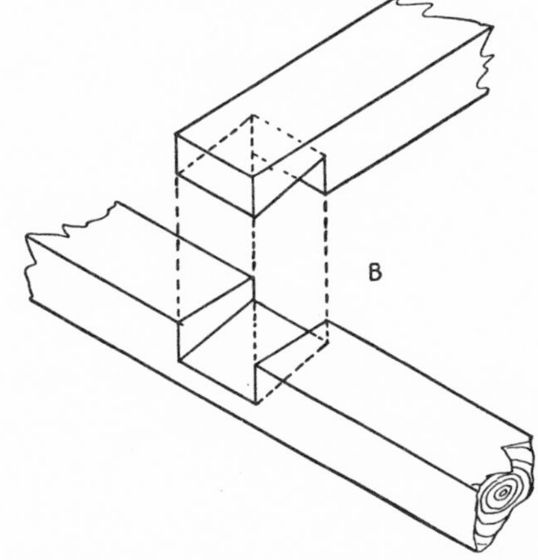

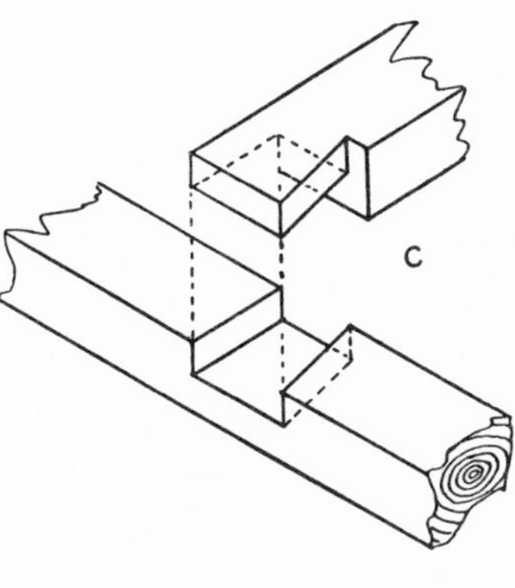

Fig. 615.—Halving A, plain; B, bevelled: C, dovetailed

respectively. A more elaborate form of halving for two pieces crossing each other is given in fig. 616. These joints may be further secured with spikes, screws, trenails, &c. In fig. 615 the two members form a right angle; such joints are known as "transverse halving". In the case of (say) two wall-plates in the same straight line, a similar joint formed at the meeting of the two pieces is known as "longitudinal halving".

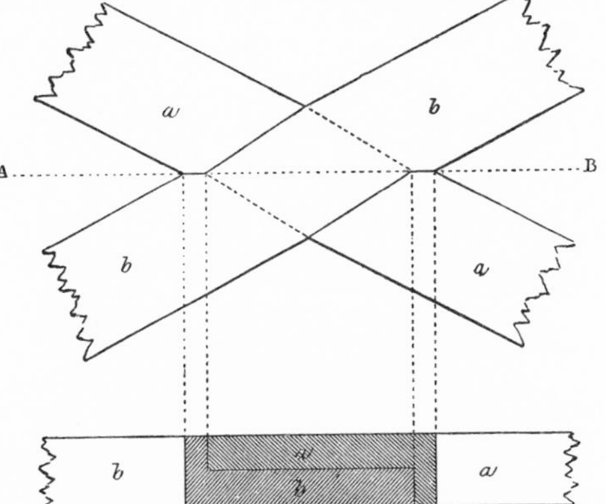

Fig. 616.—Halving at Intersection of two Timbers

Where the beam and its bearer are of the same depth and in the same plane, as in the case of a floor-joist carried by a trimmer, different kinds of joint are used. The simplest form is an ordinary mortise and tenon, the tenon on the end of the joist passing quite through the trimmer. The strength of the joint will be the smaller of the two resistances—(1) the resistance of the tenon to shearing across the fibres, and (2) the resistance of the wood below the mortise to shearing or transverse rupture. Another form of joint which may be used is dovetailed housing, similar to fig. 610, No. 1, but with the housing extending only three-fourths of the depth of the trimmer, the lowest fourth of the dovetail being cut away to correspond.

A modification of the tenon joint, known as the *tusk tenon* (fig. 617), is considered to be the best. The joint is often proportioned as follows:—If the depth of the beam or joist is divided into 12 parts, the 2 middle parts will represent the depth of the tenon, 5 parts being above the tenon and 5 below; the shoulder or "tusk" is placed at the middle of the

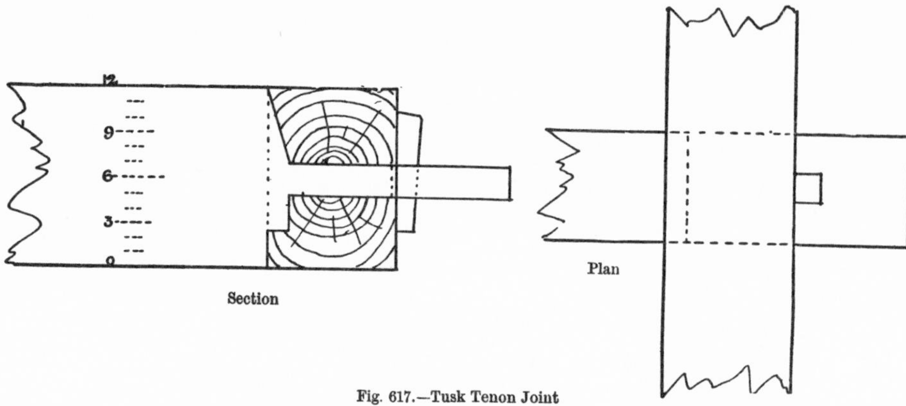

Fig. 617.—Tusk Tenon Joint

5 lower parts. In the illustration the tenon is continued through the bearing beam or trimmer, and the joint is drawn tight by means of a key or wedge driven gently through a hole in the tenon close to the back of the bearer or trimmer. If the tenon is required to resist considerable tension it must be continued 6 inches or more beyond the back of the trimmer, otherwise it will easily shear from the key outwards. In this joint the tusk bears the greater part of the load, and the shoulder above the tenon is sloped back so that neither the joist nor the trimmer is unduly weakened.

In the case of a beam supported by a post at some point in the latter other than the end, a mortise-and-tenon joint is often used, and in this case the tenon must be as large as

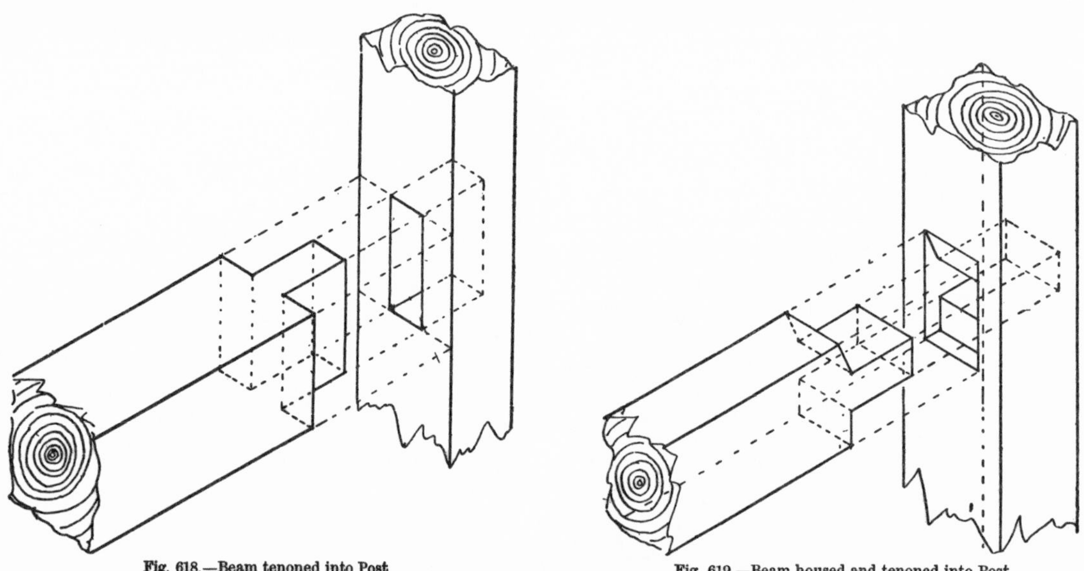

Fig. 618.—Beam tenoned into Post

Fig. 619.—Beam housed and tenoned into Post

possible without unduly weakening the post. If the same factor of safety is to be used for the beam and the post, then the joint must be proportioned as follows:—Shearing resistance of tenon is to shearing stress in beam as compression resistance of post (at mortise) is to compression stress in post due to load above the mortise. The tenon should be placed on edge (fig. 618), so as to make the horizontal area of the mortise as small as possible. A smaller tenon can be used, and a better bearing obtained, by forming the joint as shown in fig. 619.

Joints between beams and posts are often strengthened by blocks (C C, fig. 620) placed under the ends of the beams and spiked or bolted to the posts. When these blocks are used very little cutting of the posts is required. Two methods are shown at A and B. To prevent lateral movement, the beam A should be spiked to the post or secured in some other way.

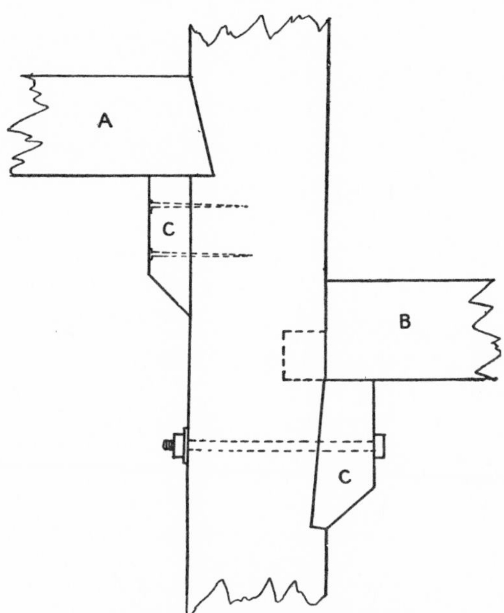

Fig. 620.—Beam and Post Joints with Blocks

When a joist has to be fixed between two beams, and it is not desired to weaken these beams in any way, a fillet of wood may be spiked or nailed to each beam, and the ends of the joists notched or grooved to fit (fig. 621). This method is often adopted for fixing ceiling-joists between tie-beams.

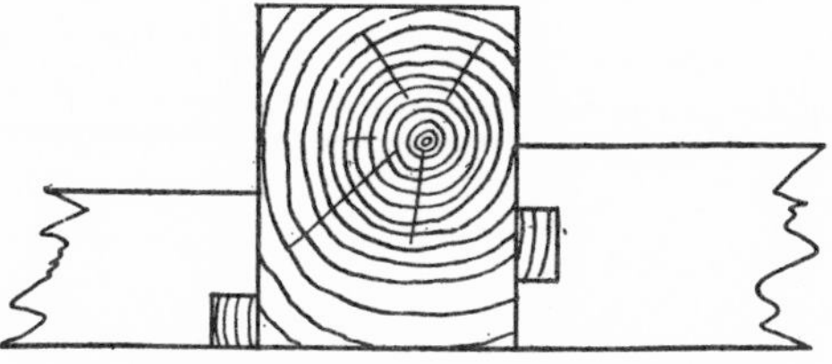
Fig. 621.—Fillet and Notch or Groove

When two beams or posts are in position, and it is required to fix a joist or other piece of timber between them, the methods shown in fig. 621 may be adopted, or an ordinary mortise-and-tenon joint may be used at one end, and a tenon and chase-mortise at the other (fig. 622). The chase-mortise is simply an ordinary mortise with a chase or groove leading to it, so that, after the first tenon has been placed in its mortise, the second tenon can be slid along the chase into position. The chase, therefore, should be an arc of a circle having the fixed tenon as centre. If the joist is between two upright posts the chase will be vertical and above the mortise.

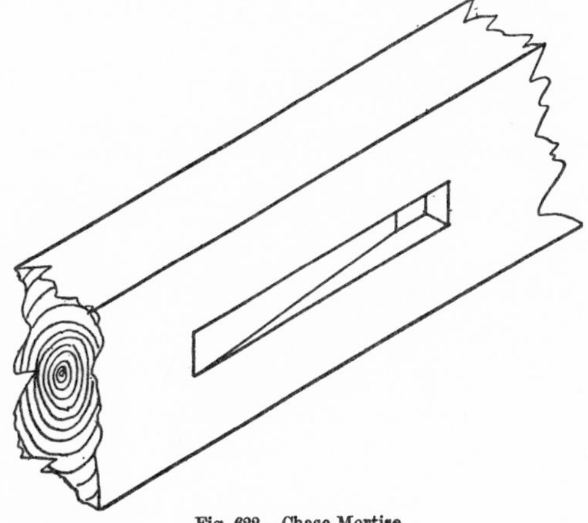
Fig. 622.—Chase-Mortise

CHAPTER II

METHODS OF LENGTHENING TIMBER

In these days of steel construction timbers of great length are not so often required as in former years, but as cases still occur, it is necessary to devote some space to the subject. The nature of the joint will differ according to the nature of the stress in the timber, and it will therefore be convenient to consider the subject under three heads: 1. Timbers in Compression, or *Posts* and *Struts*; 2. Timbers in Tension, or *Ties*; and 3. Timbers under Transverse Stress, or *Beams*.

1. *Struts.*—Timbers of this kind are not difficult to lengthen. Generally the ends of the two pieces are cut perfectly square so as to abut against each other at right angles to

the direction of the timbers, and the joint is then covered with wood or iron fish-plates or "plasters" bolted to the timbers.

Perhaps the best method is to enclose the ends in a closely-fitting iron socket; this may be a simple collar fixed to the two timbers with bolts, or may have a diaphragm across at the middle of the depth, forming a double socket, so that the iron diaphragm will come between the ends of the timbers. The latter is the better method, as the timbers obtain a better bearing and are less likely to split.

A simple scarfed joint (fig. 623) is often used, the two pieces being secured with bolts. The two overlaps forming the joint ought to be of exactly equal length, so that each bears its share of the load. Sometimes checking is adopted, as at *a* and *b*, to prevent lateral movement; but this increases the difficulty of making the joint so that all the surfaces take their fair share of the load.

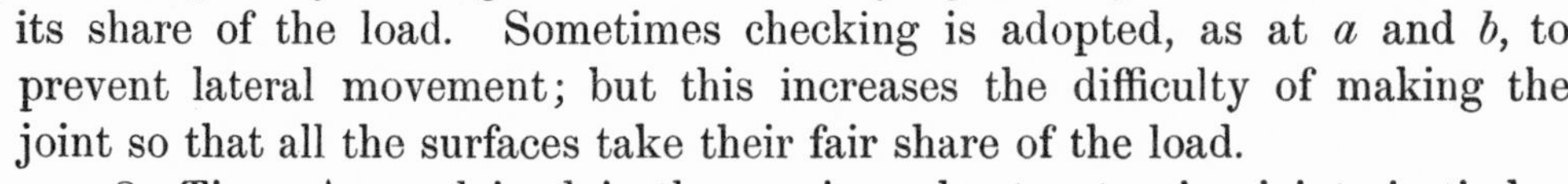

Fig. 623.—Scarfed Compression-joint

2. *Ties.*—As explained in the previous chapter, tension joints in timber are somewhat difficult to make, and almost invariably require some kind of metal fastening. Timber ties are generally lengthened either by *fishing* or *scarfing*.

Fished joints are those in which the ends of the timbers are cut square and tied to each other by means of fish-plates of wood or iron bolted to them on opposite sides. The simplest form of fished joint is shown in fig. 624. It is obvious that, for the joint to be of the same strength as the timbers, the two fish-plates must together have the same sectional area as the tie itself, if the material is the same in both cases. Thus, two fish-plates, each measuring 12 inches by 3 inches, will be required for forming the joint in a 12-inch by 6-inch tie. The length of the fish-plate and the number of bolts will depend upon the amount of the tensile stress, and upon the shearing and compressive resistance of the timber and the shearing resistance of the bolts.

To design a joint properly, every possible method of failure must be taken into consideration. The following table has been prepared to show the proportions which the several parts of the joint must bear to each other in order that none of them may be loaded either above or below its working stress.[1]

TABLE I.—WORKING STRESSES AND PROPORTIONATE AREAS OF WROUGHT-IRON AND PITCH-PINE IN FISHED JOINTS

No.		Ultimate Resistance in lbs. per square inch.	Factors of Safety.	Working Resistance in lbs. per square inch.	Proportionate Areas.					
					1.	2.	3.	4.	5.	6.
1.	Wrought-iron, Tension ...	60,000	5	12,000	1	$\frac{1}{3}$	$\frac{5}{6}$	$\frac{1}{8}$	$\frac{1}{80}$	$\frac{1}{12}$
2.	" Shearing[2] ...	40,000	10	4,000[2]	3	1	$2\frac{1}{2}$	$\frac{3}{8}$	$\frac{3}{80}$	$\frac{1}{4}$
3.	" Compression	50,000	5	10,000	$1\frac{1}{5}$	$\frac{2}{5}$	1	$\frac{3}{20}$	$\frac{3}{200}$	$\frac{1}{10}$
4.	Pitch-pine, Tension	15,000	10	1,500	8	$2\frac{2}{3}$	$6\frac{2}{3}$	1	$\frac{1}{10}$	$\frac{2}{3}$
5.	" Shearing ...	600	4	150	80	$26\frac{2}{3}$	$66\frac{2}{3}$	10	1	$6\frac{2}{3}$
6.	" Compression ...	5,000	5	1,000	12	4	10	$1\frac{1}{2}$	$\frac{3}{20}$	1

As an example of the method of using this table, we will consider a fished joint in a pitch-pine tie having a tensile stress of 108,000 lbs. If the fish-plates are of pitch-pine, the

[1] For the ultimate resistance of timber, see Section VI, Part II—Tension and Shearing, Chapter II; Compression, Chapter III; Factors of Safety, Chapter V.

[2] Calculations are generally based on the shearing resistance, but in practice bolts in timber often fail by bending. This is particularly liable to occur if the bolts become slack in consequence of the shrinkage of the wood. Experiments seem to show that for 1-inch bolts the working resistance is only 3000 lbs., or about 3800 lbs. per square inch, and for $1\frac{1}{8}$-inch bolts 4200 lbs. The sectional area of a $1\frac{1}{8}$-inch bolt is practically 1 square inch. In the table the figure 4000 has been adopted for convenience, but it must not be forgotten that this over-estimates the resistance of bolts less than $1\frac{1}{16}$ inch in diameter, and under-estimates the resistance of larger bolts.

lines relating to the tension and compression of wrought-iron will not be required. The calculations may be commenced at any one of the four remaining factors. We will begin with the bolts.

$$\text{Shearing area of bolts} = \frac{108{,}000}{4000} = 27 \text{ square inches.}$$

In order to increase the area of the timber exposed to compression, it is better to use a larger number of small bolts than a small number of larger bolts. This will be understood by comparing four 1-inch bolts with one 2-inch bolt; the four 1-inch bolts will have the same sectional area and nearly the same shearing resistance as the 2-inch bolt, but their united diameters will be twice as great, and the compression due to them will be distributed therefore over twice the area of wood. We will assume that 12 bolts must be used, arranged in two rows, each bolt being of course in double shear, and that the bolts are placed horizontally.

$$\text{Cross-sectional area of each bolt} = \frac{27}{2 \times 12} = 1{\cdot}125 \text{ square inch.}$$

$$\text{Diameter of bolt} = \sqrt{\frac{1{\cdot}125}{{\cdot}7854}} = 1{\cdot}2 \text{ inch.}$$

We have seen that the total shearing area of the bolts is 27 square inches. According to column 2 of the Proportionate Areas, the tension, shearing, and compression areas of the wood must be $2\frac{2}{3}$, $26\frac{2}{3}$, and 4 times as much respectively, that is to say—

Tension area of pitch-pine		$= 27 \times 2\frac{2}{3} = 72$	square inches.
Shearing	„	$= 27 \times 26\frac{2}{3} = 720$	„
Compression	„	$= 27 \times 4 = 108$	„

It will be best to consider the compression first. As there are 12 bolts, each 1·2 inch in diameter, the breadth of the timber can be easily found.

$$\text{Breadth of timber} = \frac{108}{12 \times 1{\cdot}2} = 7{\cdot}5 \text{ inches.}$$

The tension area of the timber must be 72 square inches, but to ascertain the scantling we must add to this the amount of wood cut out for one pair of bolt-holes. The breadth of the timber has already been found to be 7·5 inches.

$$\text{Depth of timber} = \frac{72}{7{\cdot}5} + (2 \times 1{\cdot}2) = 12 \text{ inches.}$$

As there are 2 rows of bolts, there are 4 planes of shearing. The length of each plane will be found by dividing the total shearing area by the breadth of the beam and by 4, and adding the diameters of 5 bolt-holes.

$$\text{Length of shearing plane} = \frac{720}{4 \times 7{\cdot}5} + (5 \times 1{\cdot}2) = 30 \text{ inches.}$$

This joint is shown in fig. 624. The fish-plates, being of the same material as the ties, must together be of the same sectional area.

The scantling of the tie is 12 inches by 7·5 inches, and the transverse area is reduced at the joint by two bolt-holes, each 1·2 inch by 7·5 inches, from which we can calculate that the strength of the joint is 80 per cent of the strength of the uncut portion of the tie.

Theoretically (in the case of a tie) it does not matter whether the fish-plates are at the top and bottom or at the sides, but as there is a certain amount of transverse stress due to the weight of the tie itself, it is usually better to put the fish-plates at the top and bottom.

Certain empirical rules for ascertaining the sizes of bolts, fish-plates, &c., are given in some books on carpentry, but they cannot be relied on. The only safe method is to consider every case in detail, as different results will be obtained according to the intensity of the stress, the number of bolts, &c.

The proportions shown in fig. 624 are very different from those given in works on carpentry and adopted in practice, but this is largely due to the fact that tension members are generally loaded far below the tensile working stress of the greatest section. Thus, a tie-beam 9 inches deep may be cut out to a depth of 3 inches to receive the foot of the principal rafter, and the effective section is therefore only two-thirds of the scantling, and the shearing resistance of the end of the tie-beam will probably be much less than the tensile resistance even of the smaller area. If a tie of the dimensions shown in the illustration were subjected to a stress of only 54,000 lbs. instead of 108,000 lbs., the number and diameter of the bolts could be reduced, and other alterations made, which would bring the joint to more ordinary proportions.

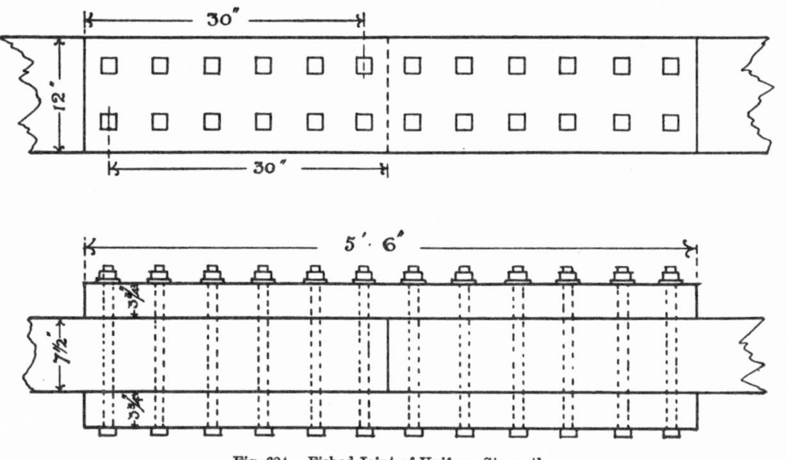

Fig. 624.—Fished Joint of Uniform Strength

The holes for the bolts must be accurately bored through the fish-plates and ties, which should be temporarily clamped or spiked together so that the holes will be exactly opposite each other. The holes should be slightly less than the diameter of the bolts, so that these will have a firm bearing when driven home. Washers should be placed under the heads and nuts of the bolts. When the bolts are screwed tight, the friction between the ties and fish-plates often gives a valuable addition to the strength of the joint, but this is not generally taken into account in the calculations, as the friction may be lost by the shrinkage of the timber. It is very important that bolts used in fished and scarfed joints, and indeed in all timber joints, should be tightened at intervals until the timber has shrunk to its ultimate dimensions.

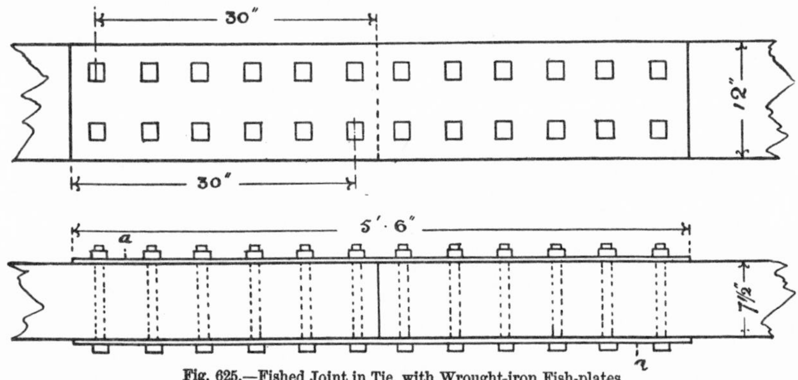

Fig. 625.—Fished Joint in Tie, with Wrought-iron Fish-plates

If wrought-iron fish-plates are used (fig. 625), the number and diameter of the bolts must obviously be the same as in the case of wood, and the position of the extreme bolts must also be the same in order to provide sufficient shearing resistance in the ties. The length of the fish-plates must therefore be the same. The thickness only remains to be calculated.

According to column 4, Table I, the tension area of wrought-iron must be one-eighth that of pitch-pine. If the depth of the fish-plates is to be the same as that of the tie, and as the bolt-holes are the same in both cases, it follows that the thickness of the fish-plates must be one-eighth that of the tie, and as there are two fish-plates, we get—

$$\text{Thickness of each wrought-iron fish-plate} = \frac{7{\cdot}5}{2 \times 8} = \tfrac{15}{32} \text{ inch, or (say) } \tfrac{1}{2} \text{ inch.}$$

It will be found that metal of this thickness will have a slightly higher compressive resistance than that of the ties, and a very much higher shearing resistance, and fish-plates ½-inch thick will therefore be strong enough in every respect.

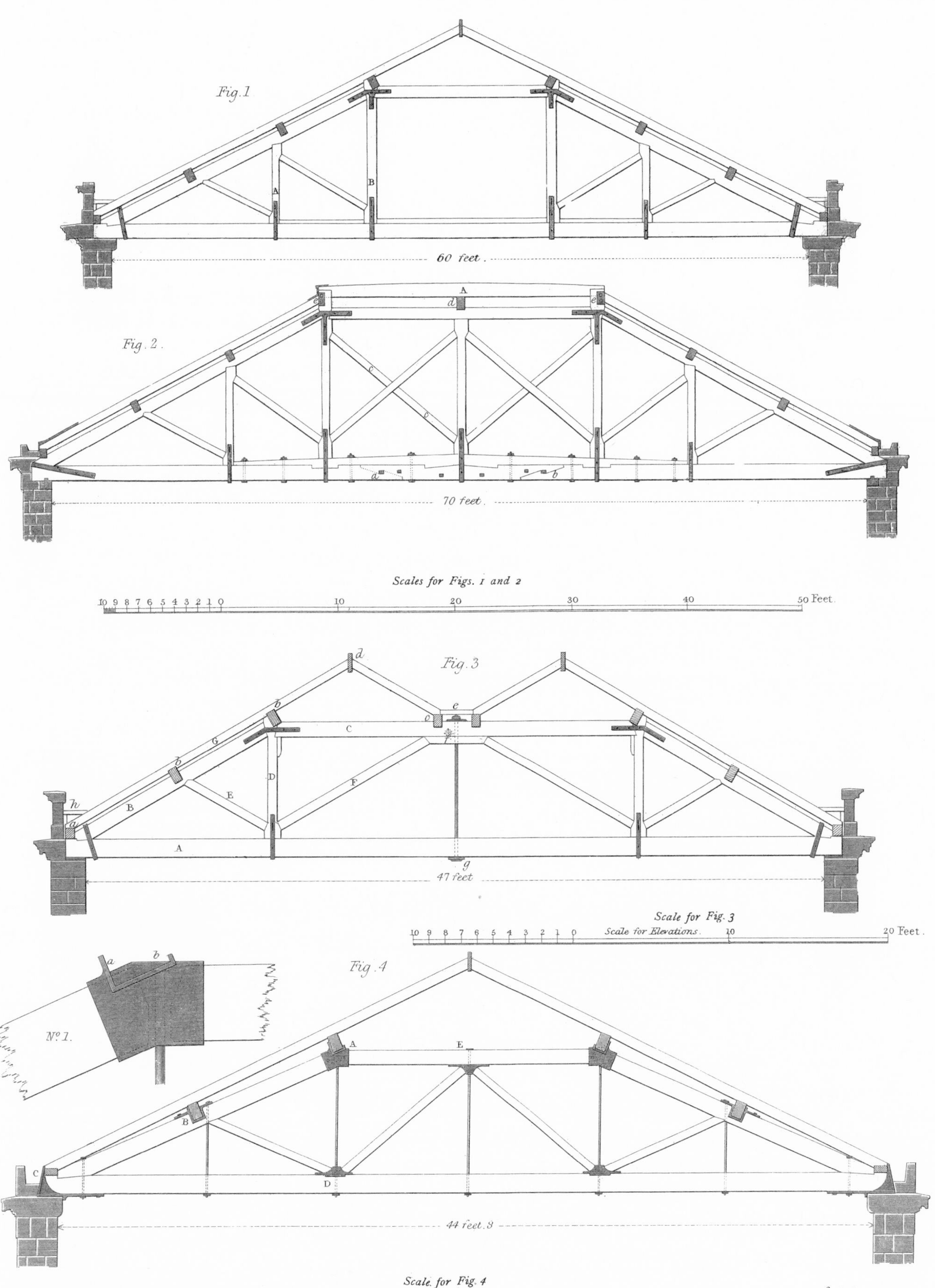

COMPOUND QUEEN-POST TRUSSES

PLATE XXVIII

COMPOUND QUEEN-POST TRUSSES

Fig. 1.—Elevation of Timber Truss containing two Queen-posts and two Princess-posts.

Fig. 2.—Elevation of Compound Timber Truss for Platform-roof, with five suspension-posts.

Fig. 3.—Elevation of Queen-post Truss with central wrought-iron suspension-rod for M-shaped roof.

Fig. 4.—Elevation of Compound Queen-post Truss with five wrought-iron suspension-rods.

No. 1: Detail of Cast-iron Socket and Purlin-rest at A.

These illustrations are more fully described on page 40, Volume II (Divisional-Volume V).

In order to reduce the length of the iron, the fish-plates are sometimes stopped at the points marked *a a* in fig. 625. The position of the extreme bolts will be the same, in order to provide the necessary shearing resistance in the ties. The bolts, however, must be slightly thicker, as the two extreme bolts at each end are in single shear only. This is equivalent to having 11 bolts instead of 12, and the thickness must be proportionately increased. This will slightly modify the dimensions of the other parts of the joint. There is no appreciable advantage in this arrangement of the plates.

In order to relieve the bolts, the fish-plates and ties are often notched to receive hard-wood joggles or keys, as shown in fig. 626, where *a a* are the ties and *b b* the fish-plates, the four keys being shaded. These keys reduce the sectional areas of the ties and fish-plates, and this reduction must be allowed for in determining their scantling. It will be observed that the keys tend to shear both the ties and fish-plates, and thus bring into play greater areas of shearing resistance than can be obtained by the bolts alone. Shorter fish-plates can therefore be used, and the bolts can be more closely spaced. Unless, however, the keys are of metal or hard cross-grained wood of large size, they will themselves fail by shearing, and in practice it is best to allow for no more addition to the shearing area than is furnished by the keys themselves.

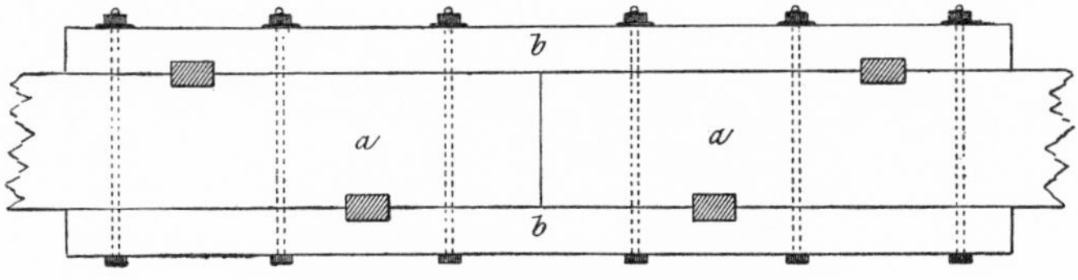

Fig. 626.—Fished and Keyed Joint in Tie

Another method of relieving the shearing stress on the bolts is to table or joggle the fish-plates into the ties. Fig. 627 shows such a joint, the parts being wedged together and united by bolts. An extreme example of this is given in fig. 628, the timbers being designed to transmit the whole of the stress, and the bolts being introduced merely to keep the fish-plates

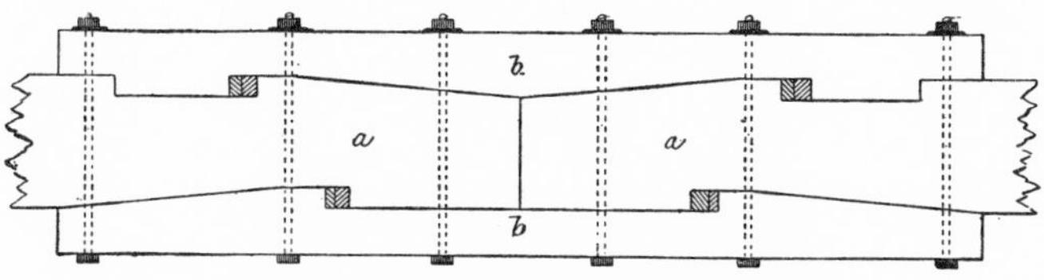

Fig. 627.—Fished, Tabled, and Keyed Joint in Tie

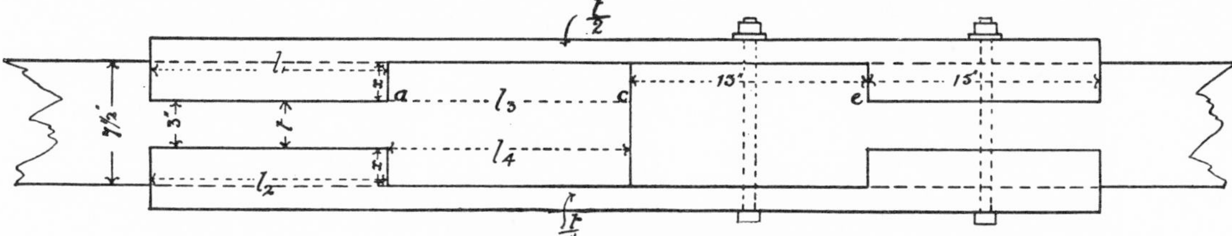

Fig. 628.—Fished and Tabled Joint in Tie

in position. We will assume that the tie has a stress of 27,000 lbs., and that all the wood is pitch-pine. The stresses which have to be considered are (1) the tension in the tie distributed over the thickness t; (2) the tension in each fish-plate distributed over the thickness $\frac{t}{2}$; (3) the shearing in the tie at l_3 and l_4; (4) the shearing in the fish-plates at l_1 and l_2; and (5) the compression on the shoulders of the tie and fish-plates at $x\,x$. We will assume that the breadth of the beam is to be 6 inches. Then (see Table I)—

$$t = \frac{27{,}000}{1500 \times 6} = 3 \text{ inches.}$$

The depth or thickness of each fish-plate must be half this amount.

$$\frac{t}{2} = \frac{3}{2} = 1\tfrac{1}{2} \text{ inch.}$$

$$l_1 = l_2 = l_3 = l_4 = 1\tfrac{1}{2} \text{ inch} \times 10 = 15 \text{ inches (see column 4).}$$

$$x = 1\tfrac{1}{2} \text{ inch} \times 1\tfrac{1}{2} = 2\tfrac{1}{4} \text{ inches (see column 4).}$$

The total depth of the tie $= t + x + x = 3 + 2\frac{1}{4} + 2\frac{1}{4} = 7\frac{1}{2}$ inches.

And the total depth of the fish-plate $= \frac{t}{2} + x = 1\frac{1}{2} + 2\frac{1}{4} = 3\frac{3}{4}$ inches.

The scantling of the tie is therefore 6 inches × $7\frac{1}{2}$ inches, while the effective scantling is only 6 inches × 3 inches (namely, at *t*). The strength of the joint is consequently only 40 per cent of the strength of the uncut portion of the tie. This shows the loss involved in making tension-joints in timber without the aid of metal fastenings. The efficiency of the joint in fig. 624 was 80 per cent.

Better results can be obtained by increasing the number of the joggles, so as to provide the requisite compression areas without cutting deeply either into the tie or fish-plates, but it is very difficult to make the joint so accurately that every shoulder receives its fair share of stress. Joints of this kind are shown in fig. 629; the notches from A to B are of the ordinary form, while those from B to C are triangular in shape.

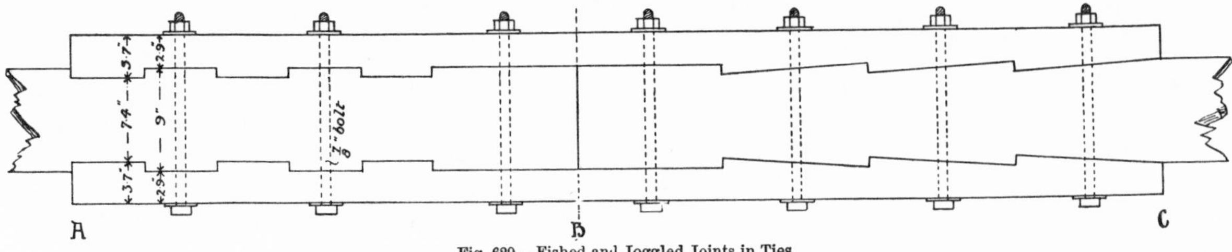

Fig. 629.—Fished and Joggled Joints in Ties

The joint shown in fig. 630 is a form of tension joint frequently used in Australia in the bottom chords of timber truss bridges. The chord is built up of four pieces, each $4\frac{1}{2}$ inches wide and 14 inches deep, overlapping each other as shown. The joint is designed for a maximum tensile stress of 109·1 tons, the working stress of the iron bolts being taken at about 4200 lbs. per square inch. The 40 bolts at the joint are $1\frac{1}{8}$ inch in diameter. The bridge for which this joint was designed has a span of 90 feet, and the truss is divided into eight bays, the two end bays being 13 feet 6 inches long, and the others 10 feet 6 inches. Joints are made in the chord in the third and sixth bays, and in the remaining bays 1-inch bolts, arranged in zigzag fashion and spaced about 1 foot apart, are used for binding the four timbers together.

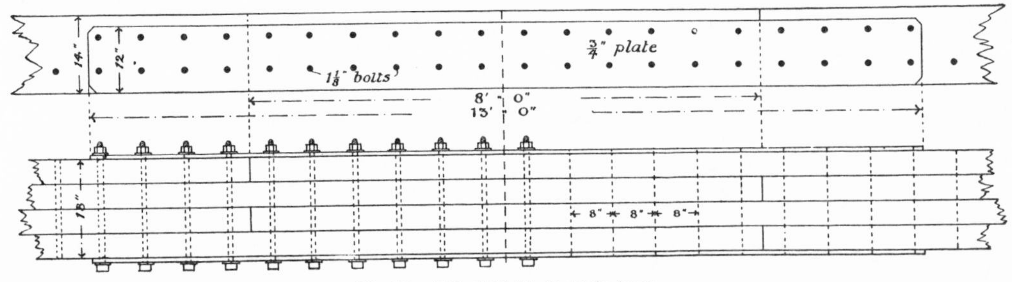

Fig. 630.—Fished Joint in Built Tie-beam

The ends of wrought-iron fish-plates are often bent at right angles and let into the timber so as to bring into action greater areas for resistance to compression and shearing. The notches cut in the timber reduce the effective area, and must be allowed for in determining the scantling.

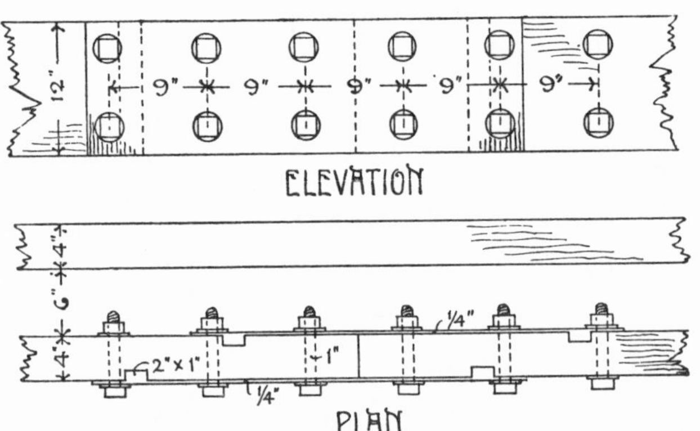

Fig. 631.—Fished Joint in Chord of Bridge Truss, with Joggled Iron Fish-plates

The joint shown in fig. 631 is a modification, the fish-plates having solid strips across them. This joint was designed for the bottom chord of a bridge truss having two 12-inch × 4-inch ties spaced 6 inches apart and a tensile stress of 53 tons. The ends of the timbers are cut square, and are united by 12-inch × $\frac{1}{4}$-inch fish-plates with 2-inch × 1-inch strips a short distance from each end. The strips are not placed opposite each other, as this would have reduced the sectional area of the wood too much. There are six 1-inch bolts on each side of the joint, four being in double shear and two in single shear. This and the previous example are taken from Professor Warren's *Engineering Construction in Iron, Steel, and Timber.*

Scarfed joints are those in which the ends of the timbers do not simply butt against each other, but are cut to overlap. Resistance to tension is obtained either by keys or joggles. Three examples are given in fig. 632. Scarfed joints involve loss of length in the timbers on account of the overlap, and are very much weaker than fished joints.

Fig. 632.—Scarfed Joints in Ties

Fished scarfs are scarfed joints with fish-plates in addition. The example in fig. 633 is from the main sill of a framed bridge trestle 100 feet high on the Oregon and Washington Territory Railroad.

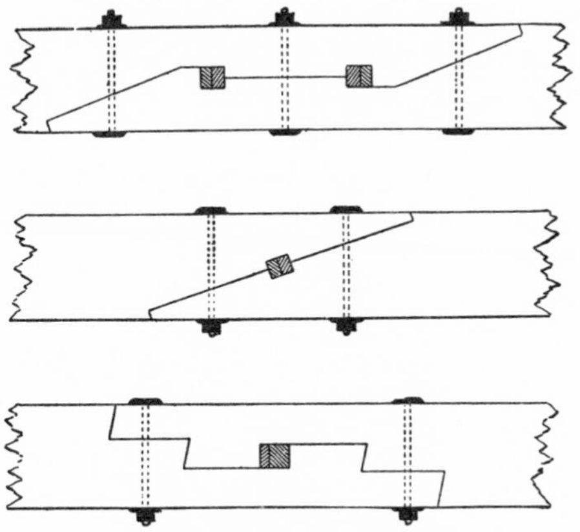

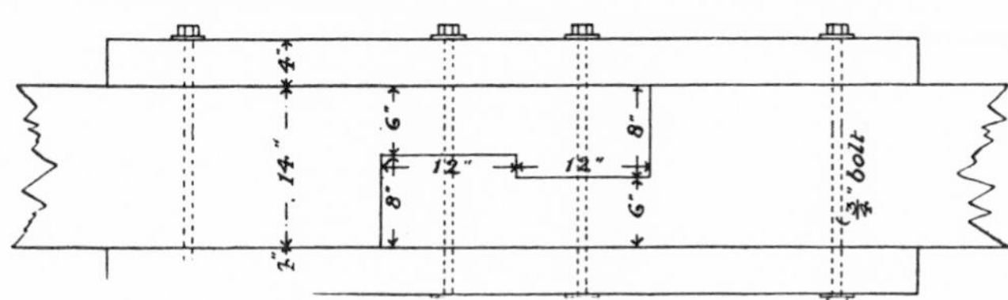

Fig. 633.—Fished-scarf Joint in Main Sill of Bridge Trestle

The sill is 14 inches deep and 12 inches wide, and the fish-plates are placed on the top and bottom and measure 4 inches × 12 inches × 6 feet. The bolts are $\frac{3}{4}$ inch in diameter. Two other examples of fished-scarf joints are given in fig. 634. Some of the joints in figs. 632 and 634 are too intricate to be of much practical use. It is almost impossible to make complicated joints so exactly that every shoulder takes its fair share of the stress.

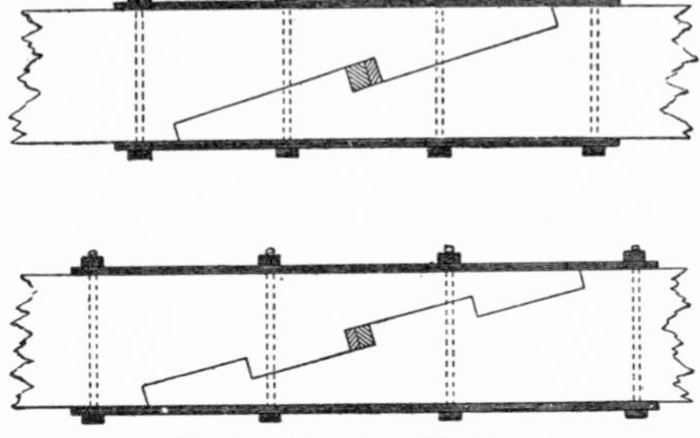
Fig. 634.—Fished-scarf Joints

3. *Beams.*—Joints in beams are more difficult to make than in ties and struts. As such joints will seldom or never be required in cantilevers, we need only consider the joints in ordinary beams. In these, assuming them to be supported at the ends and not fixed, the upper part will be in compression and the lower part in tension, but the intensity of the stresses will vary according to the loading and the position of the joint. Thus, in a beam loaded at the centre, the bending moment is greatest at the centre and diminishes uniformly to nothing at each support, and in a beam uniformly loaded the bending moment is greatest at the centre and nothing at each support, but when drawn graphically the diminution is shown by a parabolic curve and not by a straight line. In either case, however, the greatest bending moment is at the centre of the span, and there is no bending moment at all at the support. It follows, therefore, that under such circumstances the joint should be as far from the centre of the beam and as near the support as possible.

In the case of beams fixed at the ends or continuous over several supports, there are points of contraflexure at some distance from the supports (varying in the end and intermediate spans and according to the loading, &c.), and as there is no bending moment at these points, the joints ought to be made as near them as possible. In the part of the beam between the fixed end (or intermediate support) and the point of contraflexure, the *lower* portion will be in compression and the *upper* in tension. If, therefore, the joint in a beam is properly located, it need not be made as strong as the beam itself.

In ordinary cases the upper part of a beam is in compression, and the joint at this part ought to be a simple butt joint, as at *a* in fig. 635, the surfaces being at right angles to the thrust. The lower part of the beam is in tension, and a fish-plate is generally bolted to the soffit to give the necessary resistance; an oblique joint in the timber assists in stiffening the joint, as shown at *b*. Additional tensile resistance will be obtained by bending the ends

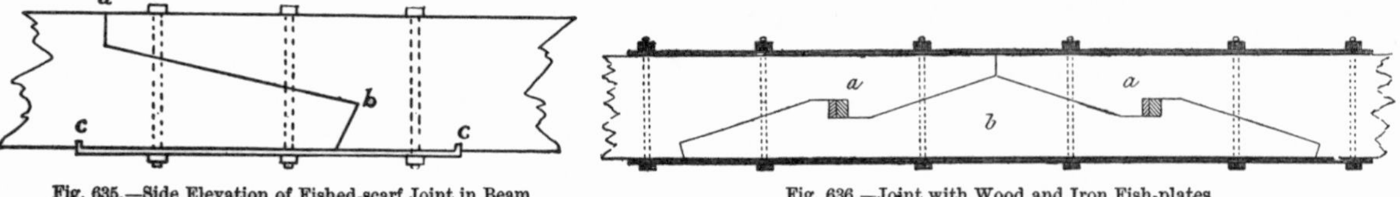

Fig. 635.—Side Elevation of Fished-scarf Joint in Beam

Fig. 636.—Joint with Wood and Iron Fish-plates

of the fish-plate and letting them into notches in the beam, as shown at *c c*, or by using a fish-plate with back strips, as shown in fig. 631. Fig. 636 gives a form of fished joint with a wood fish-plate indented into the beams, and with iron fish-plates in addition.

If a simple scarfed joint is used, the surfaces of the scarf ought to be parallel to the direction of the bending stress—in other words, parallel to the *depth* of the beam. Such a

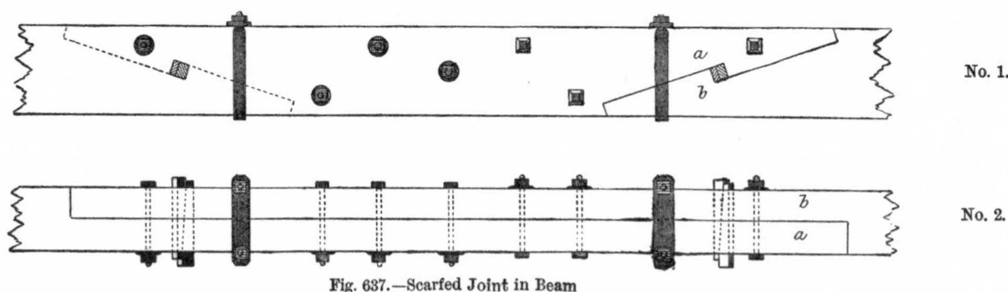

Fig. 637.—Scarfed Joint in Beam

joint is shown in fig. 637, where No. 1 is the elevation and No. 2 the plan; the joint is faulty in having the compression surfaces at the end of each scarf too small, and not at right angles to the thrust.

CHAPTER III

COMPOUND TIMBER BEAMS

It is frequently impossible to obtain a timber beam of sufficient scantling to carry safely the load which it will have to bear. In such cases two or more beams may be superposed and fixed together to form one beam of great depth; this is known as a "built" beam. Sometimes sufficient *strength* can be obtained by placing two or more timbers side by side,

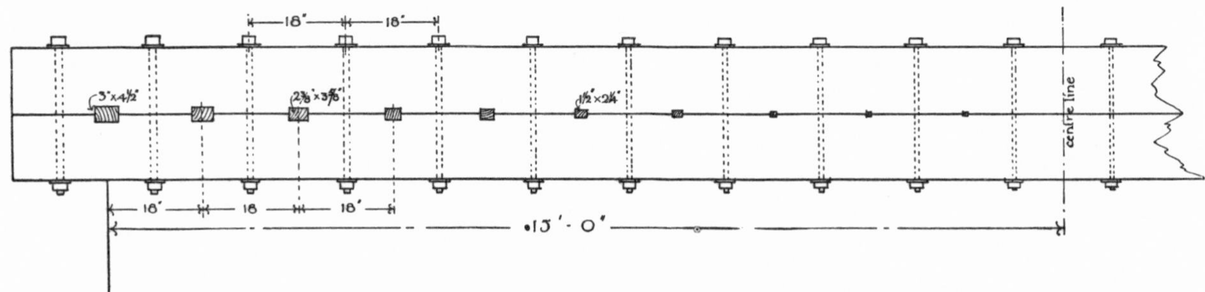

Fig. 638.—Built Beam of Two Timbers

but when the *deflection* is taken into consideration it may be found that the result is thoroughly unsatisfactory, and, if square timbers are used, the method involves a great waste of material. Or two beams may be placed side by side with an iron plate between, the whole being bolted together; this is a "flitched" beam. A third method consists in fixing tie-rods and struts under or over the beam to form a truss; the beam is then said to be "trussed".

1. *Built Beams.*—These generally consist of two beams placed one above the other, bolted together, and prevented from sliding upon each other by hardwood keys, as shown in fig. 638. As an example of this method of construction, we will consider the case of a Baltic fir or red-deal beam required to carry a load of 24,000 lbs. equally distributed, the span being 30 feet, and the ends of the beam supported and not fixed. The modulus of transverse rupture of sound red deal is about 6000 inch-lbs., and we will assume the safe working stress to be 1000.

The usual formula for this case will be adopted, namely, $w = \frac{4fbd^2}{3l}$. Transposing this we have $bd^2 = \frac{3lw}{4f}$.

$$\text{Therefore, } bd^2 = \frac{3 \times 360 \text{ inches} \times 24{,}000 \text{ lbs.}}{4 \times 1000 \text{ inch-lbs.}} = 6480 \text{ inches.}$$

If the available beams are square, d will be equal to $2b$.

$$\text{Therefore, } b \times (2b)^2 = 4b^3 = 6480 \text{ inches,}$$

$$b^3 = \frac{6480}{4} = 1620 \text{ inches, and } b = \sqrt[3]{1620} = 11{\cdot}75 \text{ inches.}$$

The beams must therefore be $11\frac{3}{4}$ inches square, or (say) 12 inches.

To determine the size of the keys (which we will suppose to be spaced 18 inches from centre to centre), we must ascertain the shearing stresses.

The maximum stress will occur at each support, and the maximum intensity per square inch will be found by the formula, $s = \frac{3w}{4bd}$.

$$\text{Maximum shearing stress per square inch} = \frac{3 \times 24{,}000}{4 \times 12 \times 24} = 62{\cdot}5 \text{ lbs.}$$

As the load is equally distributed, the shearing stress will decrease regularly towards the centre of the span, where it will be zero, but it will be well to consider each 18-inch space between the wedges as loaded uniformly with the *maximum* shearing stress in that space. The greatest intensity of shearing stress in the ten spaces from each support to the centre will be as follows:—

No.				
No. 1	...	62·5	to 56·25	lbs. per square inch.
„ 2	...	56·25	„ 50	„ „
„ 3	...	50	„ 43·75	„ „
„ 4	...	43·75	„ 37·5	„ „
„ 5	...	37·5	„ 31·25	„ „
No. 6	...	31·25	to 25	lbs. per square inch.
„ 7	...	25	„ 18·75	„ „
„ 8	...	18·75	„ 12·5	„ „
„ 9	...	12·5	„ 6·25	„ „
„ 10	...	6·25	„ 0	„ „

The total stresses in the spaces, along the plane of greatest intensity (taking the maximum as the basis of calculation), will be—

Total shearing stress (No. 1) = 12 inches × 18 inches × 62·5 lbs. = 13,500 lbs.
„ „ („ 2) = 12 „ × 18 „ × 56·25 „ = 12,150 „
„ „ („ 3) = 12 „ × 18 „ × 50 „ = 10,800 „

and so on, being 1350 lbs. less in each space as the centre is approached.

The length of the keys will be regarded as equal to the breadth of the beam, namely 12 inches. Let x = the breadth of key required. The safe shearing resistance of the keys, which are of British oak, is taken at 250 lbs. per square inch. Then—

$$\text{Shearing area of key No. 1} = 12x,$$

$$\text{and } 12x \times 250 \text{ lbs.} = 13{,}500 \text{ lbs.}$$

$$\therefore x \text{ (No. 1)} = \frac{13{,}500}{12 \times 250} = 4{\cdot}5 \text{ inches.}$$

$$\text{Similarly, } x \text{ („ 2)} = \frac{12{,}150}{12 \times 250} = 4{\cdot}05 \text{ „}$$

$$x \text{ („ 3)} = \frac{10{,}800}{12 \times 250} = 3{\cdot}6 \text{ „}$$

and so on, the breadth of the keys diminishing ·45 inch for each space towards the centre.

The bolts will be variably stressed exactly in the same manner. Taking the safe tensile resistance of wrought-iron at 12,000 lbs. per square inch, we can proportion them as follows:—

$$\text{Sectional area of No. 1 bolt} = \frac{13{,}500}{12{,}000} = 1{\cdot}125 \text{ square inch (or } 1\tfrac{1}{4} \text{ inch in diameter),}$$

$$\text{,, \quad ,, 2 ,,} = \frac{12{,}150}{12{,}000} = 1{\cdot}0125 \text{ ,, \quad (,, } 1\tfrac{3}{16} \text{ ,, \quad ,,),}$$

$$\text{,, \quad ,, 3 ,,} = \frac{10{,}800}{12{,}000} = 1{\cdot}9 \text{ ,, \quad (,, } 1\tfrac{1}{8} \text{ ,, \quad ,,),}$$

and so on.

The thickness of the keys remains to be considered.

The total shearing stress on No. 1 wedge is 18,000 lbs., and the breadth of the notch on which this force acts is 12 inches. Taking the safe resistance of red deal to fibre-crushing at 750 lbs. per square inch, we have—

$$\text{Depth of notch No. 1} = \frac{13{,}500}{12 \times 750} = 1{\cdot}5 \text{ inch.}$$

$$\text{,, \quad ,, 2} = \frac{12{,}150}{12 \times 750} = 1{\cdot}35 \text{ ,,}$$

$$\text{,, \quad ,, 3} = \frac{10{,}800}{12 \times 750} = 1{\cdot}2 \text{ ,,}$$

SIZE OF KEYS

	Depth.	Breadth.		Depth.	Breadth.
No. 1	3 inches	4·5 inches.	No. 6	1·5 inches	2·25 inches.
,, 2	2·7 ,,	4·05 ,,	,, 7	1·2 ,,	1·8 ,,
,, 3	2·4 ,,	3·6 ,,	,, 8	·9 ,,	1·35 ,,
,, 4	2·1 ,,	3·15 ,,	,, 9	·6 ,,	·9 ,,
,, 5	1·8 ,,	2·7 ,,	,, 10	·3 ,,	·45 ,,

If the beams are placed close together, the thickness of each key will be of course twice the depth of the notch, but frequently a space of 2 inches or more is allowed between the beams for ventilation, and the thickness of the keys is increased by the same amount. This, however, subjects the keys to a transverse stress, and their breadth must be increased accordingly.

As the variation in the sizes of the keys and bolts renders the execution of the work more difficult, the *spaces* may be increased from the support towards the centre so that the

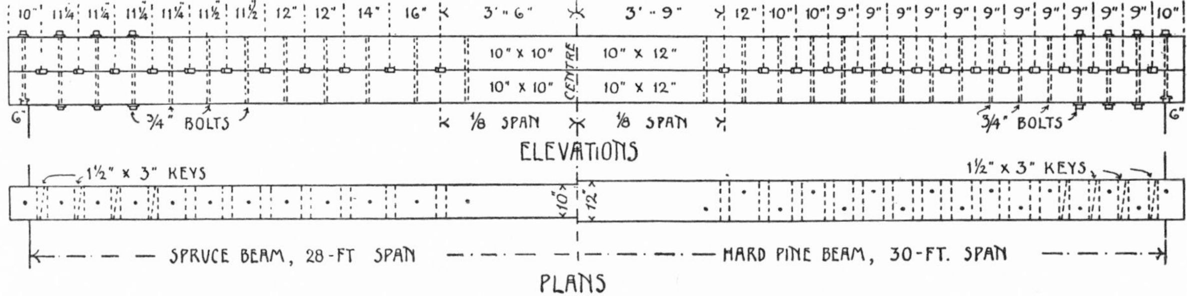

Fig. 639.—Keyed and Bolted Compound Beams of Spruce and Hard Pine

stresses in the several spaces are approximately equal, and the keys and bolts can therefore be all of the same size.

In the case of beams loaded only at the centre, and supported at the ends, the shearing stress is uniform throughout the length of the beam (except at the very centre of the span, where it is 0). Consequently the spaces, keys, and bolts will be uniform throughout the beam; but it must not be forgotten that a beam loaded at the centre will bear only half as much as a similar beam with a distributed load.

Mr. F. E. Kidder in his *Building Construction and Superintendence* mentions a series of tests of compound beams, carried out by Prof. Edgar Kidwell of the Michigan College of Mines in 1896–97. The arrangements shown in fig. 639 were found to give the best results. The strength of a spruce beam keyed and bolted as shown, with oak keys, is said to be 95 per cent of that of a solid beam of the same size, but the deflection is from 20 to 25 per cent more than that of a solid beam. If cast-iron keys are used, the deflection is reduced almost to the normal amount. The keys should be wedge-shaped, as shown in fig. 640, so that they can be driven to a firm bearing against the sides of the notches.

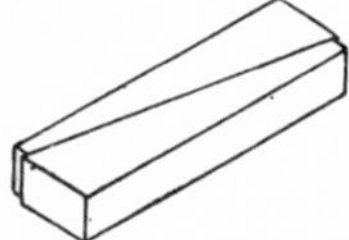

Fig. 640.—View of Key for Compound Beams

Mr. Kidder states that a 20-inch × 10-inch spruce beam composed of two 10-inch × 10-inch timbers, keyed and bolted as in the left-hand portion of fig. 639, will bear safely a distributed load of 15,000 lbs. for a span of 28 feet, and that a 20-inch × 12-inch beam of hard pine (such as "Georgia pine",—that is to say, "pitch-pine") composed of two 12-inch × 10-inch timbers, keyed and bolted as in the right-hand half of fig. 639, will bear safely a distributed load of 24,000 lbs. for a span of 30 feet. Beams from 16 to 20 inches deep should have 1½-inch × 3-inch keys, ¾-inch bolts and 3-inch washers; beams 24 inches deep should have 2-inch × 4-inch keys, ⅞-inch bolts and 3½-inch washers; and beams 28 inches deep 2¼-inch × 4½-inch keys, ⅞-inch bolts and 3½-inch washers. For beams of greater width than 10 inches, the bolts should be staggered as shown in the right-hand half of the illustration. It is said that 1½-inch × 3-inch keys should be *at least* 11¼ inches from centre to centre for spruce and white pine, and 9 inches for Oregon and Georgia pines, 2-inch × 4-inch keys 15 inches and 11½ inches, and 2¼-inch × 4½-inch keys 17 inches and 13 inches. If the spacing of the keys works out at less than these distances, the safe load must be reduced or the breadth of the beam increased. The number of keys on each side of the centre is regulated by the depth of the beam and the kind of timber; spruce beams 20 inches deep must have 11 keys on each side and pitch-pine 15; for 24-inch beams the numbers must be 9 and 13, and for 28-inch beams 10 and 14.

Beams 20 inches deep of spruce or white pine may be used for spans of 24 to 32 feet, and similar beams of hard pine up to 36 feet; 28-inch beams may be used for spans of 30 to 40 feet. The central portion of the beam is not keyed, the first key on each side of the centre being placed at a distance from the centre equal to one-eighth of the span.

For beams loaded at the centre, the spacing of the keys should, it is said, be uniform from the first key on each side of the central space to the supports, and the safe central load will be one-half of the safe distributed load.

Another form of compound beam often used in America is shown in fig. 641. The timbers are simply superposed, and fastened together by 1¼-inch boards nailed with tenpenny nails to the two sides; the boards are laid diagonally, those on one side being at right angles to those on the other. Prof. Kidwell's tests showed that, "in every case, long before the beam broke, the struts split open, or the nails were drawn partly out or bent over in the wood, thereby permitting the component beams to slide on each other; it was found that no amount of nailing could prevent this". The efficiency of a beam of this kind is only about two-thirds of that of a solid beam of the same size (not including the thickness of the boarding).

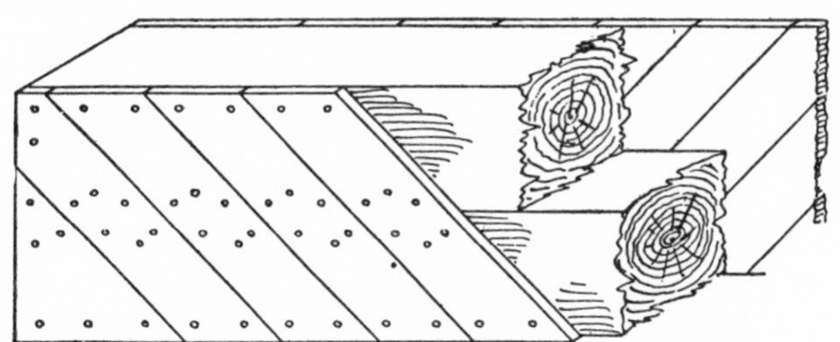

Fig. 641.—Compound Beam with Diagonal Boarding at Sides

Sometimes three timbers are combined to form one beam, either by being placed on the top of each other, or by being arranged as shown in fig. 642. The latter arrangement is somewhat wasteful, as also is the arrangement, sometimes adopted, of four timbers forming a square.

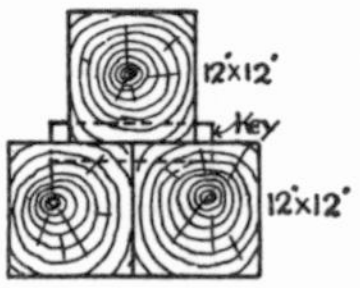

Fig. 642.—Built Beam of Three Timbers

For American railway bridges, two 12-inch × 12-inch timbers on the top of each other are considered sufficient for spans of 12 to 15 feet, three (arranged as in fig. 642) for spans of 16 to 20 feet, and four (arranged in a square) for spans of 21 to 25 feet. In Australia built beams composed of three 12-inch × 12-inch timbers have been used for road bridges up to 42 feet span.

In built beams of the kind illustrated in figs. 638 and 639 the timber is disposed to the best advantage, and the great proportionate depth ensures considerable stiffness.

Sometimes the beams are indented into each other throughout their length, after the manner of scarfed joints; but this method involves great labour and also a waste of material, and cannot therefore be recommended.

2. *Flitched Beams.*—The ordinary flitched beam consists of two timbers placed side by side with a wrought-iron plate between them, the whole being secured with wrought-iron bolts about 18 inches apart arranged in zigzag fashion (technically, "staggered"). Flitched beams are often of service in wood floors where the depth of the bridging joists must not be exceeded, as, for example, in the case of trimmers or trimming joists which may be called upon to bear heavy loads. They are also useful for the main beams of floors and as bressummers, &c., although they are not as frequently used now as before the days of rolled iron and steel joists. Large flitched beams are generally made by sawing a square log longitudinally down the centre line, and turning the two pieces so that the butt of one and the top of the other come together, and so that the sawn faces are exposed. This method has the advantage of revealing any defects in the heart of the log, and also hastens the seasoning of the timber.

Sometimes a single piece of timber is used with two flitch-plates, one on each side. This is known as a double-flitched beam. It is an arrangement which cannot be recommended for new structures, as the timber is not so well ventilated and therefore more liable to decay; but it may be of service for strengthening an existing beam which has proved too weak for its work.

It is commonly said that the thickness of the wrought-iron plate in a single-flitched beam should be about one-twelfth the breadth of the two timbers, in order that the timber and iron may be stressed in the same proportion with regard to their ultimate strength. In other words, the ultimate resistance of the timber is assumed to be one-twelfth of that of the iron. Of course the ratio will vary according to the nature of the timber and iron.

The only experiments (as far as I know) which have been made on flitched beams were carried out at Woolwich Arsenal in 1859. They were not by any means exhaustive, but they showed clearly that the flitch increased both the strength and stiffness of the beam to a very considerable extent. The tests are recorded in Table II. The moduli of rupture and of transverse elasticity have been worked out from the formulas given in Chapter III, Part I, of the preceding section. The strength of the flitched beams (Nos. 4 and 5) is about $2\frac{1}{2}$ times as much as that of the unflitched beams (Nos. 1, 2, and 3), and the stiffness is increased almost in the same proportion.

If from the breaking load of the flitched beam No. 4 (34,862 lbs.) we deduct the breaking load of the corresponding beam without flitch (No. 1, 13,102 lbs.), the increase of strength due to the iron is found to be 21,760 lbs. Working this out by the ordinary formula for rectangular beams, we have $\mathrm{w} = f\frac{4 \times \frac{1}{2} \times 9^2}{3 \times 204} = 21{,}760$ lbs., from which we find that $f = 82{,}204$ inch-lbs.

Comparing Nos. 2 and 5 in a similar manner, we have $w = f\frac{2 \times \frac{1}{2} \times 9^2}{3 \times 204} = 11{,}279$ lbs., and $f = 85{,}219$ inch-lbs.

The average value of f is therefore about 83,700 inch-lbs., and this is not far from the true modulus of rupture of wrought-iron. Sir Benjamin Baker's experiments led him to the conclusion that for rectangular sections the modulus of rupture of wrought-iron and steel is about 70 per cent more than the ultimate resistance to tension. This would make the modulus of rupture about 102,000 lbs., but it must not be forgotten that the bolt-holes in the flitch reduce its strength, and that we have not taken these into consideration.

TABLE II.—EXPERIMENTS MADE AT THE ROYAL ARSENAL, WOOLWICH, IN 1859, ON THE STRENGTH OF MEMEL DEALS (SUPPORTED AT THE ENDS), WITH AND WITHOUT WROUGHT-IRON FLITCH-PLATES

No.	Description.	Total Scantling.		Clear Span.	How loaded.	Breaking Weight.	Modulus of Rupture (f).	Modulus of Transverse Elasticity (E).	Remarks.
		Breadth. ins.	Depth. ins.	ins.		in lbs.	in inch-lbs.	in inch-lbs.	
1.	Two deals 9 ins. by 3 ins. side by side ...	6	9	204	Uniformly	13,102	4,124	1,378,000	Broke in middle.
2.	Do.	6	9	204	At centre	6,800	4,281	898,000	Do.
3.	Do. bolted together with twelve $\frac{3}{4}$-inch wrought-iron bolts ...	6	9	204	Uniformly	13,503	4,250	1,247,000	Do.
4.	Two deals with 9 ins. by $\frac{1}{2}$ in. wrought-iron plate between, bolted together with eleven $\frac{3}{4}$-inch bolts	$6\frac{1}{2}$	9	204	„	34,862	10,130	2,689,000	Both timber and iron snapped asunder in middle.
5.	Do.	$6\frac{1}{2}$	9	204	At centre	18,079	10,507	3,065,000	Broke in middle.
6.	Do. but with iron flitch only 9 ft. 3 ins. long, and secured with ten bolts	$6\frac{1}{2}$	9	204	Uniformly	21,566	...	...	Broke at bolt-holes at one end of flitch, the iron being uninjured.
7.	Do.	$6\frac{1}{2}$	9	204	At centre	14,873	...	...	Broke at one end of iron flitch, the iron being uninjured.
8.	Single beam	9	12	204	Uniformly	27,076	3,196	853,000	Broke in two places near middle at a cluster of small knots.
9.	Do.	6	9	204	„	11,879	3,740	1,179,000	Broke near middle at a cluster of small knots.

Assuming, however, for a moment that the modulus of rupture is 102,000 lbs., the ultimate transverse strength of the flitch-plate will only be 26,470 lbs. (uniformly distributed), or 8392 lbs. less than the ultimate resistance of the flitched beam No. 4. It is clear, therefore, that both the timber and the iron are stressed by the load, and the strength of a flitched beam may be ascertained by calculating the strength of the timber and iron separately and adding the results together. Or the calculations may be performed at one operation as follows:—

1. *Flitched Beam uniformly loaded and supported at the ends,*

$$w = \frac{4\,d^2}{3\,l}(f b + f_1 b_1) \quad \ldots \quad (37)$$

2. *Flitched Beam loaded at the centre and supported at the ends,*

$$\text{W} = \frac{2\,d^2}{3\,l}\,(f\,b + f_1\,b_1) \quad \ldots \quad \ldots \quad \ldots \quad \ldots \quad \ldots \quad (38)$$

In these formulas—

W = breaking weight.
d = depth in inches, this being the same for the wood and iron.
l = length in inches.

b = breadth of timbers in inches.
b_1 = breadth of iron flitch in inches.
f = modulus of rupture for timber.
f_1 = modulus of rupture for iron.

The value of f (for red deal) according to these tests is only about 4200; this is lower than the values recorded in Table XII, page 356. It is interesting to note the still lower values obtained for the 9-inch × 6-inch and 12-inch × 9-inch beams. Probably the low results obtained at Woolwich (both for f and E) were partly due to the method of loading.

The value of f_1 (for wrought-iron) may be taken to be about 90,000.

The tests with short flitch-plates in the central portion of the beams (Nos. 6 and 7) show a considerable increase of strength over the unflitched beams (Nos. 1 to 3), but the plates were not long enough. The bending moment at the end of the plate ought not to exceed the resistance of the timber at that point; to comply with this requirement, the plate in No. 6 beam, which was uniformly loaded, should have been about 15 feet 6 inches long, and that in No. 7, which was centrally loaded, about 10 feet 6 inches long. In the case of uniformly-loaded beams, therefore, the flitch cannot be much less in length than the beam itself; but in beams with a single load a shorter plate may be used, the length depending upon the relative resistances of the iron and wood. In practice, the flitch is generally coextensive with the timber.

In bolting the beams together, care should be taken that a bolt-hole is not bored in the centre of the span, if the beam is uniformly loaded, or under the concentrated load whether this is at the centre or some other point.

3. *Trussed Beams.*—The third method of strengthening beams is by trussing them. In many cases the truss is no deeper than the beam, but the gain in strength is not worth the

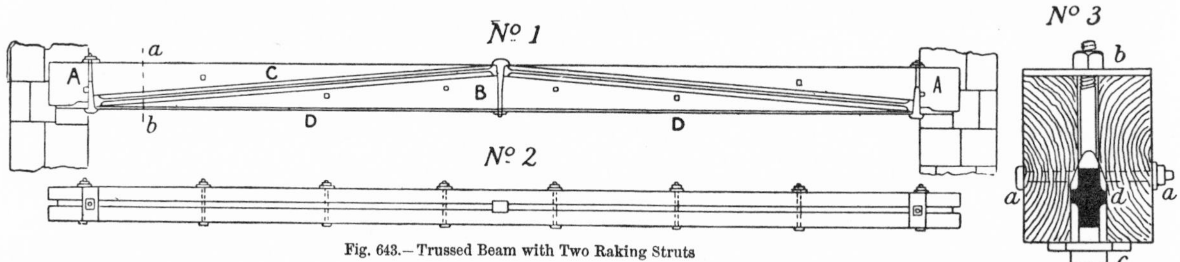

Fig. 643.—Trussed Beam with Two Raking Struts

trouble. The truss may have the apex at the top, as in fig. 643, or at the bottom, as in fig. 644. Trusses of the former kind have two raking struts, while those of the latter kind have two raking ties. Fig. 643 is given as an example of trussing within the depth of

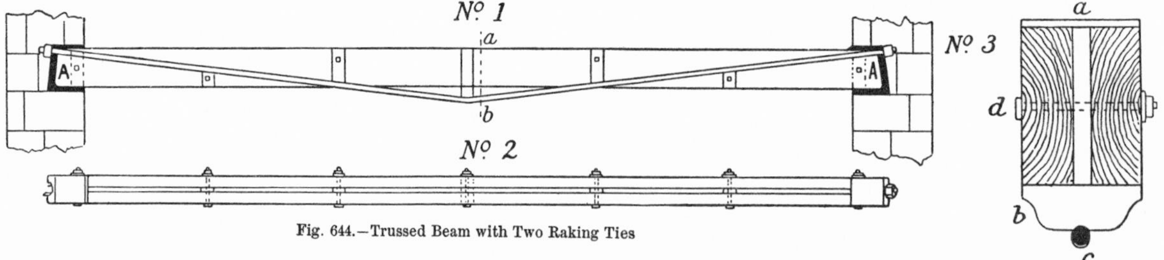

Fig. 644.—Trussed Beam with Two Raking Ties

the beam, but cannot be recommended. No. 1 is an elevation with one of the flitches removed to show the trussing. No. 2 is a plan of the beam, and No. 3 a section through the line *a b*. The trussing-bars C, No. 1, are of cast-iron, and are shown in section enlarged at *d*, No. 3. An iron tension-plate D extends along the bottom of the beam, and connects the abutment bolts A A. These bolts pass between the flitches, and are screwed down upon an iron plate *b*. The central bolt B fulfils the functions of the king-post of a trussed roof. The beam is generally sawn in two, and the ends reversed, when put together in a truss.

Fig. 644 is an example of a girder trussed with a short strut or stirrup-piece *b*, end-plates A A, and a tension-rod A *b* A. No. 1 is an elevation of the beam; No. 2 a plan; and No. 3 an enlarged vertical section through the line *a b*. It is impossible to tighten the tension-rod after it is fixed, and for this purpose a right-and-left union joint may with advantage be made in the rod near the point *b*. The stirrup-piece is too short to be of much service, and it is better to use a cast-iron strut of the form shown in fig. 645. The web above the cap-plate is placed between the two flitches of the beam and secured to them with bolts. A solid oak strut is sometimes used, with a tenon passing between the two flitches.

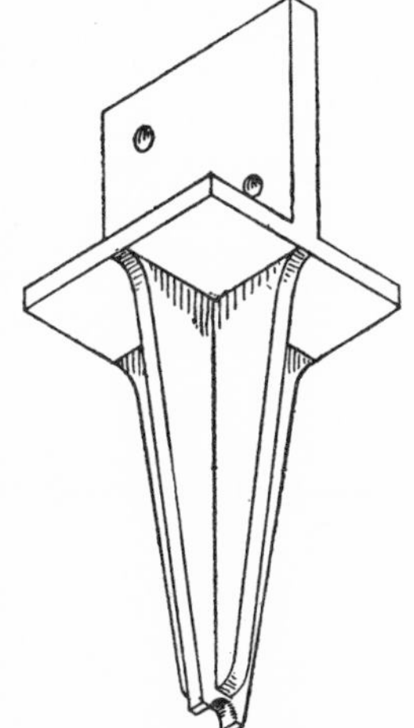

Fig. 645.—Cast-iron Strut for Trussed Double Beam

For spans exceeding about 20 feet two cast-iron struts A A are used, as shown in fig. 646, and these ought to be as long as possible, as the quicker rake of the tie-rods increases the strength of the truss. In this example the beam is solid, and each end of the tie-rod is passed through a hole bored in the beam, the latter being protected from injury by an iron chair B. The hole at its commencement in the end of the beam ought to have its centre at about one-fourth the depth of the beam. A right-and-left union joint is shown at C; this may take the form of a screwed link. In the case of large trusses additional members are sometimes introduced at D E, particularly if the beam is subjected to unsymmetrical or rolling loads.

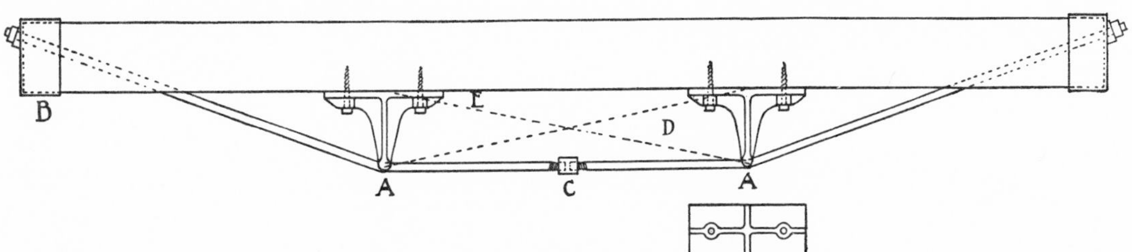

Fig. 646.—Trussed Beam with Tie-rod and Two Struts

The depth of the truss, measured from the centre of the beam to the centre of the tie-rod, is generally made about $\frac{1}{10}$ the span of the beam.

Sometimes two tie-rods are used, one on each side of the beam, as shown in fig. 647, the ends passing through ears or lugs cast on the chair at the end of the beam. Or the beam may be composed of three flitches with spaces between for the two rods.

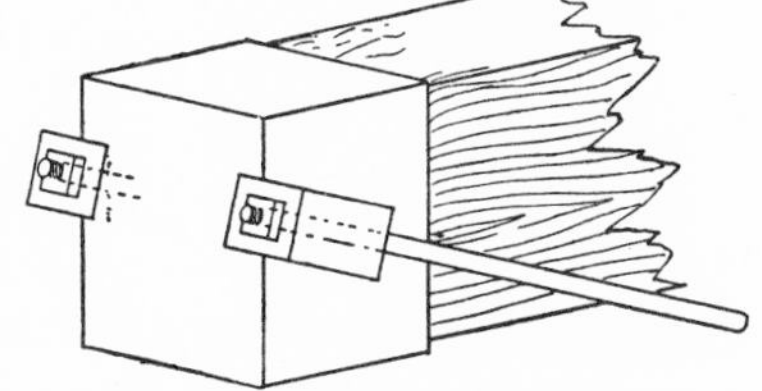

Fig. 647.—Cast-iron Chair for Two Tie-rods

In calculating the strength of a trussed girder, it is generally assumed that the truss must be strong enough to bear the whole of the load without any allowance for the transverse strength of the timber. In the case of a king-post truss (fig. 648), the stresses can easily be found graphically. Let it be required to find the stresses in a trussed beam, 30 feet span, and loaded with 20,000 lbs. equally distributed, the truss to be 3 feet deep.

The load must be considered as concentrated at the three joints, namely, A B = $\frac{W}{4}$, B C = $\frac{W}{2}$, and C D = $\frac{W}{4}$. The reaction at each support will be $\frac{W}{2}$. The stress diagram is shown in No. 2, from which the stresses in the several numbers can be read by scale, namely B F and

C G (compression) 15,000 lbs., F E and G E (tension) about 15,800 lbs., and F G (compression) 10,000 lbs.

With a working tensile resistance of 12,000 lbs., the sectional area of each tie-rod will be $\frac{15,800}{12,000} = 1{\cdot}317$ square inch (or nearly $1\frac{5}{16}$ inch in diameter).

With a working compressive resistance of 7,000 lbs. (for cast-iron), the sectional area of

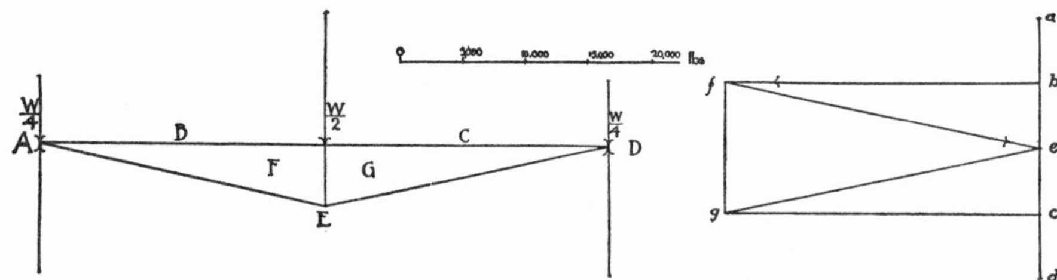

Fig. 648.—Stresses in Trussed Beam

the strut must be $\frac{10,000}{7,000} = 1\frac{3}{7}$ square inch. In practice this would be largely exceeded; the plate abutting against the timber must be of sufficient area to prevent indentation.

With a working compressive resistance of 1000 lbs. (for pitch-pine), the sectional area required to resist the compression due to the truss will be $\frac{15,000}{1,000} = 15$ square inches.

The stresses caused by the distributed load between the three points of support remain to be considered. For this purpose the half-beam (15 feet long) may be regarded as supported at the outer end and fixed at the centre, and it has a load of 10,000 lbs. uniformly distributed. The formula for this case is $= \text{W}\frac{4 f b d^2}{3 l}$.

For pitch-pine $f = 1200$ inch-lbs. (allowing a factor of safety of about 6).

Transposing the formula, we have $b d^2 = \frac{3 \text{w} l}{4 f}$.

$$\text{Therefore, } b d^2 = \frac{3 \times 10,000 \times 180}{4 \times 1200} = 1125.$$

If $b = d$, $d^3 = 1125$, and $d = \sqrt[3]{1125} = 10{\cdot}4$ inches.

A beam $10\frac{1}{2}$ inches square will therefore be strong enough, but as it would be deficient in stiffness, a deeper beam will be better. Let $d = 2b$. Then $4b^3 = 1125$, and $b = \sqrt[3]{\frac{1125}{4}} = 6{\cdot}5$ inches, and $d = 2b = 13$ inches.

To this must be added a sectional area of 15 square inches to resist the compression due to the trussing. This may be done by increasing the size of the beam to 13 inches by $7\frac{3}{4}$ inches, or $15\frac{1}{2}$ inches by $6\frac{1}{2}$ inches. In practice a beam 14 inches by 7 inches would be used, as two can be cut out of one square timber.

The determination of the strength of a queen-post trussed beam presents no difficulties when the load is symmetrical, and will be easily accomplished on the lines indicated above and by reference to Case 8, Chapter IV, p. 322. The advantage of the queen-post truss is that it divides the beam into three parts and so reduces the transverse stress. Thus, in the case of a beam similar in all respects to that shown in fig. 648, but with a queen-post truss instead of a king-post truss, the length of each bay will be 10 feet instead of 15 feet, and the load on each bay will be $\frac{20,000}{3} = 6666{\cdot}\dot{6}$ lbs. Therefore,

$$b d^2 = \frac{3 \times 6666{\cdot}\dot{6} \times 120}{4 \times 1200} = 500.$$

Then if $d = 2b$,

$$4b^3 = 500, \text{ and } b^3 = \frac{500}{4} = 125,$$

$$\therefore b = \sqrt[3]{125} = 5 \text{ inches, and } d = 2b = 10 \text{ inches.}$$

A beam 5 inches broad and 10 inches deep will therefore suffice, instead of one $6\frac{1}{2}$ inches by 13 inches. To this must of course be added the sectional area required to resist the compression caused by the trussing.

In the case of a trussed beam supporting a floor, the joists ought to rest on the top of the beam or be carried by stirrup irons, in order that the beam may not be weakened by notching.

An interesting example of trussing above the beam is given in fig. 649. It was designed by Mr. F. E. Kidder for a clear span of 18 feet and a distributed load of 36,000 lbs. The beam consists of two flitches, each 3 inches broad and 14 inches deep; the struts are 6 inches by 10 inches, and the vertical tie-rods $1\frac{3}{8}$ inch in diameter. The girders are

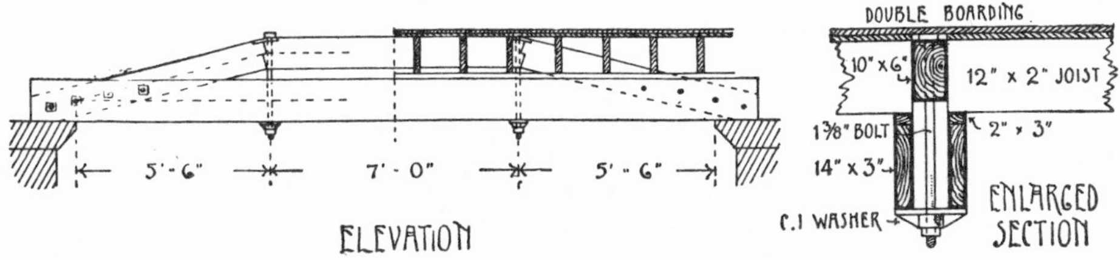

Fig. 649.—Beam with Queen-post Truss having Wood Struts and Iron Tie-rods

spaced 16 feet from centre to centre, and the floor-joists (12 inches by 2 inches) rest on 2-inch by 3-inch packings above the flitches. The timber is Georgia pine. Heavy cast-iron plates or washers are fixed under the beam to receive the lower ends of the tie-rods, and cast- or wrought-iron plates, made to fit the angles of the truss, are fixed to receive the upper ends. The foot of each raking strut passes between the two flitches of the beam, and is secured with four $1\frac{1}{2}$-inch bolts. This arrangement is suitable for rooms where trusses below the beams would be unsightly, or would reduce the effective height to an inconvenient degree.

CHAPTER IV

CARPENTERS' IRONWORK

Details have already been given of some of the ironwork commonly used in timber structures, and other details will be given in subsequent chapters; but it will be useful to gather into a separate chapter some information about the carpenters' ironwork in common use.

TABLE III.—PROPERTIES OF IRON AND STEEL

	Average weight per cubic ft. in lbs.	Safe Tensile Strength in lbs. per square inch.	Safe Shearing Strength in lbs. per square inch.	Safe Crushing Strength in lbs. per square inch.	Safe Modulus of Transverse Rupture in lbs. per square inch.
Cast-iron	450	2,600	7,000	13,500	4,000
Wrought-iron	480	10,000	7,500	10,000	12,000
Steel, medium ...	490	12,500	7,500	13,000	16,000

Wrought-iron Bolts.—The Standard Whitworth thread is generally adopted for bolts and nuts in this country. The thread is inclined at an angle of 55°, one-sixth of the thread at top and bottom being rounded off. Seller's thread is the standard in the United States, and is inclined at an angle of 60°, one-eighth being taken off the top and bottom and left flat. In considering the strength of bolts, the area at the bottom or sunk portion of the thread must be adopted as the basis of calculation, and not the full diameter of the bolt.

The following table gives particulars of bolts and nuts from $\frac{3}{8}$ inch to 2 inches in diameter. The thickness of the nut is generally equal to the diameter of the bolt, while the thickness of the head is seven-eighths of the diameter. Bolt-heads are generally square, each side being equal to the diameter of the nut across the flats, but bolts with button heads and countersunk heads are also made as shown in fig. 650. Nuts are either square or hexagonal, the sizes being given in the table.

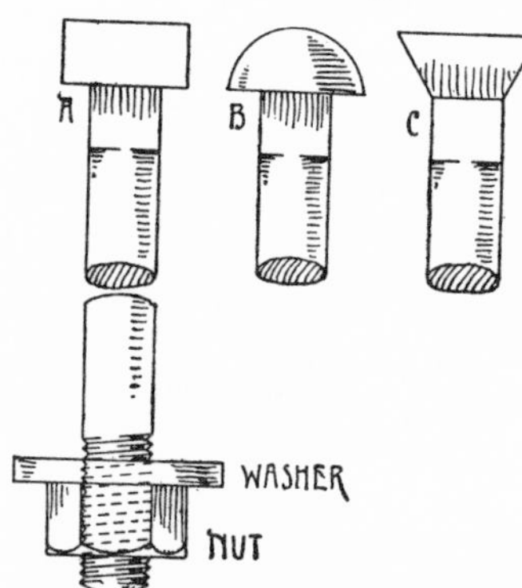

Fig. 650.—Bolt-heads, Washer, and Hexagonal Nut:—A, Square head; B, button or cup head; C, countersunk head.

Washers ought always to be used under the heads and nuts of bolts in timber structures in order to prevent indentation of the wood; the thickness should be about $\frac{1}{3}$ of the diameter of the bolt, and the diameter not less than 3 times that of the bolt. The thickness of small washers is often No. 14 B.W.G.

TABLE IV.—WHITWORTH STANDARD BOLTS AND NUTS

Diameter of Bolt and Thickness of Nut.	Number of Threads per inch.	Diameter of Bolt at Bottom of Thread.	Sectional Area at Bottom of Thread.	Thickness of Head.	Diameter of Nut across Flats.	Diameter of Nut across Corners.	Safe Tensile Loads at 6000 lbs. per square inch.	Safe Tensile Loads at 8000 lbs. per square inch.	Safe Tensile Loads at 10,000 lbs. per square inch.
ins.		in.	sq. ins.	in.	ins.	ins.			
$\frac{3}{8}$	16	·295	·068	·33	·71	·82	409	546	684
$\frac{1}{2}$	12	·393	·121	·43	·92	1·06	728	971	1,213
$\frac{5}{8}$	11	·508	·202	·54	1·10	1·27	1,216	1,621	2,228
$\frac{3}{4}$	10	·622	·303	·65	1·30	1·5	1,822	2,430	3,038
$\frac{7}{8}$	9	·733	·422	·76	1·48	1·7	2,532	3,376	4,220
1	8	·840	·554	·87	1·67	1·95	3,325	4,433	5,542
$1\frac{1}{8}$	7	·942	·697	·98	1·86	2·15	4,181	5,575	6,968
$1\frac{1}{4}$	7	1·067	·893	1·09	2·04	2·36	5,358	7,144	8,932
$1\frac{3}{8}$	6	1·161	1·057	1·20	2·21	2·55	6,342	8,456	10,570
$1\frac{1}{2}$	6	1·286	1·29	1·31	2·41	2·78	7,740	10,320	12,900
$1\frac{5}{8}$	5	1·369	1·47	1·42	2·57	2·97	8,820	11,760	14,600
$1\frac{3}{4}$	5	1·494	1·74	1·53	2·75	3·18	10,464	13,952	17,440
$1\frac{7}{8}$	$4\frac{1}{2}$	1·590	1·99	1·64	3·01	3·48	11,916	15,888	19,860
2	$4\frac{1}{2}$	1·715	2·31	1·75	3·14	3·63	13,860	18,480	23,100

Bolts, especially small ones, are often severely strained by being screwed too tightly. The holes for bolts should be bored about $\frac{1}{16}$ inch less than the diameter of the bolts, so that these will fit tightly when driven home.

When two timbers meet at right angles, spikes or drift-bolts may be used if lateral movement only has to be prevented, as in the case of posts, sills, and beams; but if the joint

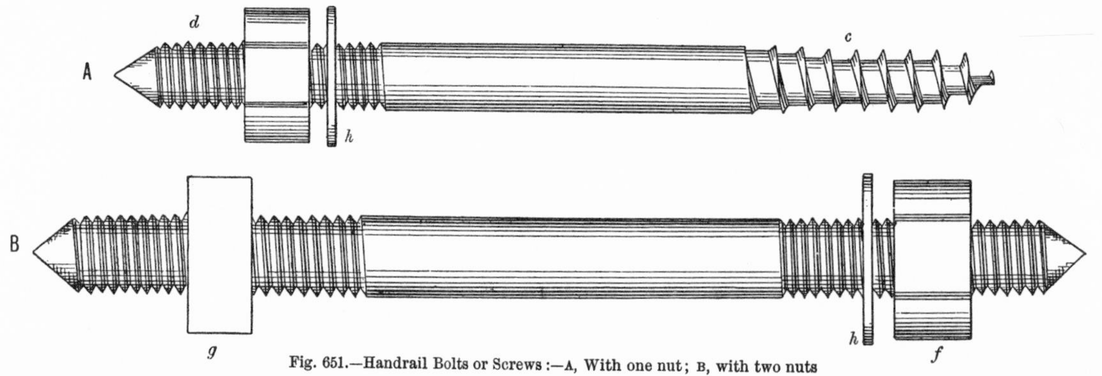

Fig. 651.—Handrail Bolts or Screws :—A, With one nut; B, with two nuts

has to transmit a tensile stress, such a method of fastening is not suitable. The joint between a king-post and tie-beam will serve as an example. A coach-screw may be driven through a hole bored through the tie-beam, and then screwed up into the king-post; but as it is screwed in the direction of the grain, its holding power is comparatively small. A better arrangement is to use an ordinary bolt and nut, the latter being inserted in a mortise cut in the king-post to receive it.

Table V.—APPROXIMATE WEIGHT OF BOLTS, INCLUDING SQUARE HEADS AND NUTS

Length under Head, in inches.	Diameter, in Inches.					Length under Head, in inches.	Diameter, in Inches.			
	$\frac{1}{2}$.	$\frac{5}{8}$.	$\frac{3}{4}$.	$\frac{7}{8}$.	1.		$\frac{5}{8}$.	$\frac{3}{4}$.	$\frac{7}{8}$.	1.
6	0·59	1·01	...	...	...	15	1·79	2·72	3·89	5·34
7	0·64	1·10	...	...	...	16	1·87	2·84	4·06	5·56
8	0·70	1·19	...	...	...	17	1·96	2·97	4·23	5·78
9	0·75	1·27	...	...	...	18	2·05	3·09	4·40	6·00
10	0·81	1·36	2·10	3·05	4·23	19	...	3·21	4·57	6·22
11	0·86	1·44	2·22	3·22	4·45	20	...	3·34	4·74	6·44
12	0·92	1·53	2·35	3·39	4·67	21	...	3·46	4·90	6·66
13	0·97	1·62	2·47	3·55	4·89	22	...	3·59	5·07	6·88
14	1·03	1·70	2·59	3·72	5·11	24	...	3·83	5·41	7·32

When two small timbers are joined end to end, "handrail" bolts or screws may be used. Two forms are shown in fig. 651. That marked A has a wood-thread screw at *c* for screwing into the end of one of the timbers, and a metal-thread screw at *d*. This end is passed into a hole of the necessary length bored in the end of the other timber, and the nut *e* is inserted in a mortise cut at a suitable point; the nut has longitudinal grooves, by means of which it can be screwed tight. The second form, marked B, has a similar nut at *f*, and a square nut at *g* for letting into a square mortise of the same size, which prevents the nut turning when the nut *f* is being tightened. Washers are inserted at *h*. These bolts are made in stock sizes from 4 to 10 inches long and from $\frac{5}{16}$ to $\frac{1}{2}$ inch in diameter.

Coach-screws with square or hexagon heads are often used for fixing timbers to iron or steel pillar-caps or brackets, and for other purposes. They are also known as lag-screws or wood-screws. The heads may be square, hexagonal, or "Gothic-headed", as shown at A, B, and C in fig. 652. They can be made of any size to suit the work in hand. Square heads are generally used in carpentry. A washer ought to be placed under the head, the diameter being three times that of the screw.

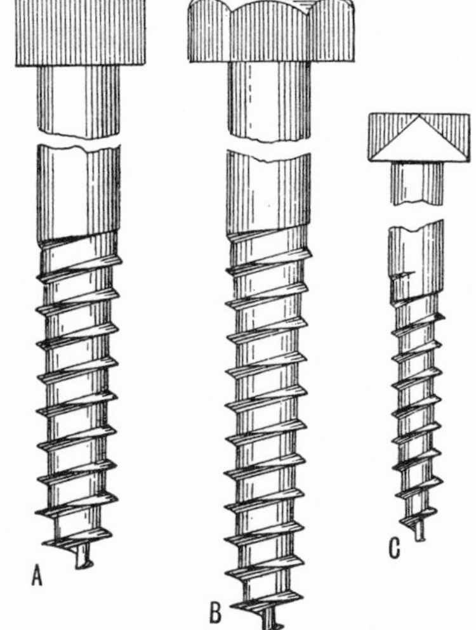

Fig. 652.—Coach-screws with Square, Hexagonal, and Gothic Heads

From Bevan's experiments on the holding power of small screws (about 2 inches long), published in 1827, the following formulas have been deduced (Hurst's edition of Tredgold's *Carpentry*):—

Holding power in hardwood $= 200{,}000\, d\,\delta\, t$ .... (39)

" softwood $= 100{,}000\, d\,\delta\, t$ .... (40)

where d = diameter of screw, δ = depth of thread, and t = the depth of penetration, all in inches.

These formulas give about five times the holding power obtained by Kirkaldy in his experiments with coach-screws, and the constants ought to be reduced to 40,000 and 20,000 respectively in order to obtain the ultimate resistance, and to 4000 and 2000 to obtain the safe working stress. Table VI contains the results of Kirkaldy's tests on the holding power of coach-screws. The tests are in two series—Nos. 1 to 4, and Nos. 5 to 7. The results are the average of three tests on each kind of timber in the first series, and of six tests in the second series. The red deal or fir used in Nos. 1 to 4 is described as "Memel", that in Nos. 5 to 7 as "Baltic".

TABLE VI.—HOLDING POWER OF COACH-SCREWS

No.	Nominal Diameter of Coach-screw.	Diameter at Bottom of Thread.	Number of Threads per inch.	Depth in Wood.	Ultimate Pulling Stress, in lbs.		
					Teak.	English Oak.	Red Deal.
	in.	in.		ins.			
1	$\frac{7}{8}$	·68	4·00	6	21,463	20,342	9901
2	$\frac{3}{4}$	·57	4·5	4	11,562	11,243	5745
3	$\frac{5}{8}$	·45	4·75	$4\frac{1}{2}$	10,827	10,274	5387
4	$\frac{1}{2}$	·42	5	3	7,922	6,724	3722
5	1	·80	3	5	...	18,625	8957
6	$\frac{7}{8}$	·69	3·5	5	...	17,514	8073
7	$\frac{3}{4}$	·63	4	5	...	16,225	7794

The holding power of coach-screws depends partly on friction, but the principal resistance is afforded by the shearing resistance of the timber. Some experiments carried out for the U.S. Government in 1874–77 showed that the resistance of "screw-bolts" to being drawn out was only about 50 per cent more than the resistance of plain round rods. Kirkaldy's tests, however, show clearly that the shearing resistance of the timber plays a very important part; the resistance afforded by teak and oak is practically the same, and is double that given by red deal. They appear to point to a simple but empirical formula—

$$w = d\,l\,s \qquad (41),$$

where w = ultimate resistance to pulling, in lbs.,
d = diameter of coach-screw, in inches,
l = length of penetration of screw, in inches,
s = shearing resistance *across* grain, in lbs. per square inch, 4000 for oak and teak, and 2000 for red deal (see page 341).

This formula, however, can only be regarded as the roughest approximation, as it does not take into consideration the nature of the thread.

Drift-bolts.—These are merely a kind of long nail, and those required in heavy timber structures are generally made from $\frac{3}{4}$-inch round or square iron with a head forged at one end and a point or chisel-edge at the other. The length of the bolt should be such as to allow it to penetrate to a sufficient depth to obtain a firm hold. Bolts 20 inches long are often used for fastening 12-inch timbers to posts, &c. The following are the approximate weights of drift-bolts 18 inches and 24 inches long:—$\frac{3}{4}$ inch round, 2·3 lbs. and 3 lbs.; $\frac{3}{4}$ inch square, 2·9 lbs. and 3·8 lbs.; 1 inch round, 4 lbs. and 5·3 lbs.; 1 inch square, 5·1 lbs. and 6·8 lbs.

Experiments show that the holding power of drift-bolts is very great. In the *Engineering News* of February 28, 1891, three series of experiments are recorded; the principal conclusions are given in the next three paragraphs, which are quoted from Foster's "Treatise on Wooden Trestle Bridges".

"*U.S. Government Experiments.*—These experiments were made under the direction of General Weitzal by Assistant U.S. Engineers A. Noble and C. P. Gilbert in 1874 and 1877, and were published by Colonel O. M. Poe in his report to the Chief of Engineers in 1884. This series was very extensive, but the valuable results obtained are robbed of much of their value by the lack (in the original publication) of suitable comparisons and conclusions. The mean of from 150 to 200 experiments with round and square bolts, both ragged and smooth, in different-sized holes, shows that the resistance after having been driven seven months is 10 per cent greater than the resistance immediately after driving, the different sizes and forms being strikingly uniform. The mean of 150 experiments under various conditions shows that the resistance to being drawn in the direction in which it was driven is only 60 per cent of its resistance to being drawn in the opposite direction; that is to say, the resistance to being drawn *through* is only 60 per cent of that to being drawn *back*. The mean of 50 experiments shows that smooth rods have a greater holding power, both to being drawn through and also to being drawn back, than ragged ones, a 'moderate' ragging reducing the resistance a little more than 25 per cent, and an 'excessive' ragging reducing the holding power more than 50 per cent.

"*Brooklyn Bridge Experiments.*—Experiments made in connection with the construction of the East River Bridge by Mr. F. Collingwood and Colonel Paine, and communicated by the former, gave a holding power of 12,000 lbs. per linear foot of bolt for a 1-inch round rod driven into a $\frac{15}{16}$-inch hole in first-quality Georgia pine, and a resistance of 15,000 lbs. in a $\frac{14}{16}$-inch hole. It was found that in lighter timber containing less pitch the holding power was about 20 per cent less; and in very dense wood, containing more pitch, about 10 per cent more.

"*University of Illinois Experiments.*—A third series of experiments was made by Mr. J. B. Tscharner in the testing laboratory of the University of Illinois, and published in full in 'No. 4, Selected Papers of the Civil Engineers' Club of the University of Illinois'. According to these experiments the average holding power of a 1-inch round rod driven into a $\frac{15}{16}$-inch hole in pine, perpendicular to the grain, is 6000 lbs. per linear foot, and under the same conditions the holding power in oak is 15,600 lbs. per linear foot. The holding power of the bolt driven parallel to the grain is almost exactly half as much as when driven perpendicular to the grain. If the holding power of a 1-inch rod in a $\frac{15}{16}$-inch hole be designated as 1, the holding power in a $\frac{14}{16}$-inch hole is 1·69; in a $\frac{13}{16}$-inch hole, 2·13; and in a $\frac{12}{16}$-inch hole, 1·09. The holding power decreases very rapidly as the bolt is withdrawn."

Captain Fraser, R.E., in vol. xxi, *R.E. Professional Papers*, gives the following summary of the results of experiments:—"In fir timber the holding power in lbs. is—For iron dogs, from 600 to 900 lbs. for each inch in length of spike; for spike nails, from 460 to 730 lbs. per inch in length, exclusive of thickness of cover-plate; for oak, ash, or beech trenails, 2000 lbs. per square inch of section; for fir (Scotch), spruce, or elm trenails, from 1000 to 2000 lbs. per square inch. The holding power of spikes in hardwood is increased approximately one-third. All the trenails were of seasoned wood."

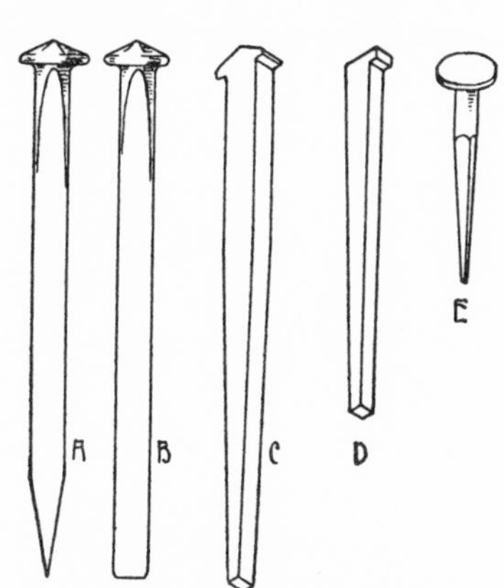

Fig. 653.—Some Varieties of Nails

A, Rose nail with sharp point; B, rose nail with flat point; C, clasp; D, brad; E, clout.

Spikes and Nails.—Spikes are a kind of large nails, and are used in carpentry in lengths from about 3 to 8 or 10 inches. They may be either *cut* spikes, which are stamped out of sheet metal, or *boat* spikes, which are forged out of square bars of wrought-iron, the latter being the stronger. Common nails are also used from "10-penny" nails (3 inches long and 60 to the lb.) to "60-penny" (6 inches long and 10 to the lb.); the intermediate sizes are 12-penny ($3\frac{1}{4}$ inches), 16-penny ($3\frac{1}{2}$ inches), 20-penny (4 inches), 30-penny ($4\frac{1}{2}$ inches), 40-penny (5 inches), and 50-penny ($5\frac{1}{2}$ inches). As the terms 10-penny, 12-penny, &c., are now nearly obsolete, and at the best are indefinite, it is always better to state the length and kind required. The length of nails should be about two-and-a-half times the thickness of the board or plank which they are intended to secure.

Boat spikes are the best for heavy timber structures. The dimensions may be as follows:—

$\frac{1}{4}$ inch square	from 3 to 5 inches long.	$\frac{1}{2}$ inch square	from 5 to 8 inches long.
$\frac{5}{16}$,,	,, 3 ,, 6 ,,	$\frac{9}{16}$,,	,, 6 ,, 10 ,,
$\frac{7}{16}$,,	,, 4 ,, $6\frac{1}{2}$,,	$\frac{5}{8}$,,	,, 8 ,, 10 ,,

Among the varieties of nails used by the carpenter may be mentioned the following:—

1. *Rose nails* (fig. 653, A and B), so called from the shape of the head; they may be wrought, cut, or pressed, and with sharp or chisel points (as at A and B), or with square ends.

2. *Clasp nails* (fig. 653, C), which have oblong heads sloping downwards from a central ridge; they may be wrought or cut.

3. *Brads* (fig. 653, D), the heads projecting on one side only; these are much used for flooring, and the smaller sizes for joinery.

4. *Clout nails* (fig. 653, E), with large flat heads (plain as in the illustration, or "counter-sunk" like the head of an ordinary screw), used for fixing sheet-metal, felt, &c., to wood; they may have sharp or chisel points.

5. *Dog nails*, with heads rather large and often hemispherical, shanks round in the upper part, and points flat, used for fixing heavy ironwork, &c.

Nails may be of cast-iron for rough work, or of malleable iron (*i.e.* cast-iron annealed), or may be hand-wrought, machine-wrought, or "cut" by machinery out of sheets of iron (the heads being afterwards stamped on). Wire nails are chiefly used in this country for packing cases, and are known as French nails or *pointes de Paris*, but in America they are used for general carpentry and joinery; the common American varieties are shown full size in fig. 654.

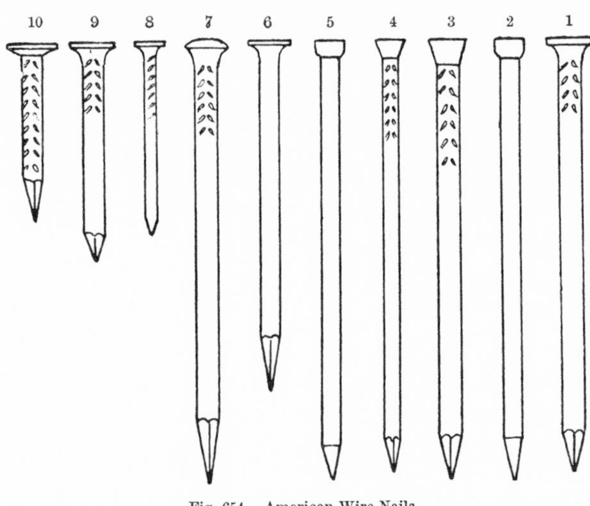

Fig. 654.—American Wire Nails

1, 8*d.* common; 2, 8*d.* common brad; 3, 8*d.* flooring brad; 4, 8*d.* casing; 5, 8*d.* finishing; 6, 6*d.* shingle nail; 7, 8*d.* clinch nail; 8, 3*d.* fine nail; 9, 3*d.* slating nail; 10, one-inch barbed roofing nail.

Mr. Bevan's experiments on the adhesive force of nails and screws in different kinds of wood were attended with the following results:—Small sprigs, 4560 in the pound, and the length of each $\frac{44}{100}$ of an inch, forced into dry Christiania deal to the depth of 0·4 inch, in a direction at right angles to the grain, required 22 lbs. to extract them. Sprigs half an inch long, 3200 in the pound, driven in the same deal to 0·4 inch depth, required 37 lbs. to extract them. Nails 618 in the pound, each nail $1\frac{1}{4}$ inch long, driven 0·5 inch deep, required 58 lbs. to extract them. Nails 2 inches long, 130 in the pound, driven 1 inch deep, took 320 lbs. Cast-iron nails, 1 inch long, 380 in the pound, driven 0·5 inch, took 72 lbs. Nails 2 inches long, 73 in the pound, driven 1 inch, took 170 lbs.; when driven $1\frac{1}{2}$ inch they took 327 lbs., and when driven 2 inches 530 lbs. The adhesion of nails driven at right angles to the grain was to force of adhesion when driven with the grain, in Christiania deal, as 2 to 1, and in green elm as 4 to 3. If the force of adhesion of a nail in Christiania deal be 170, then in similar circumstances the force for green sycamore will be 312, for dry oak 507, for dry beech 667. A common screw $\frac{1}{5}$ of an inch diameter was found to hold with a force three times greater than a nail $2\frac{1}{2}$ inches long, 73 of which weighed a pound, when both entered the same length into the wood.

Straps.—Illustrations of different kinds of wrought-iron straps and stirrups have been given in previous chapters, and little need be added here. They ought to be placed as nearly as possible in the direction of the stresses they have to sustain; for this reason the strap shown in fig. 600 is better than that in fig. 601. It is also necessary that they should be drawn tight either by gib and cotters (fig. 587) or by boring the bolt-hole at such a point that the strap will be drawn tight by driving the bolt.

END OF VOL. I